CHARLES PROTEUS
STEINMETZ

CHARLES PROTEUS STEINMETZ

CHARLES PROTEUS
STEINMETZ

A BIOGRAPHY

BY

JOHN WINTHROP HAMMOND

ILLUSTRATED

Merchant Books

David Sarnoff conducts an inspection tour of RCA Transoceanic Station at Rockymeen? ny

1921   Among Mr. Sarnoff's guests are Albert Einstein, Steinmetz, Langmuir, and other famous scientists

FOREWORD

In the last few years of his life, repeated requests came to Dr. Steinmetz for his autobiography. He always refused to write of himself, saying that it was too much trouble. Offers were made to do the work for him if he would supply the facts. Finally he decided that Mr. Hammond should be his biographer, and to him he gave many of the facts concerning himself contained in the following pages. Before his death, Dr. Steinmetz had read and approved some of the chapters contained in this volume.

Mr. Hammond has had no easy task in painting a word picture of so great and unusual a man as Dr. Steinmetz, but I feel that he has done his subject justice, and I give his work my most hearty and unsolicited commendation.

J. LeRoy Hayden.

PREFACE

A short lifetime ago, the pursuits of men and women were carried on with hardly any assistance whatever from the mysterious, hidden power of electrical energy. In 1870 no one had ever traveled in an electric car, made use of an electric elevator, or spoken to another person by telephone. True, the telegraph had ushered in one form of swift communication; but a host of other emancipating innovations waited on man's clearer knowledge of electricity and its laws.

That knowledge has come within the last fifty years. To-day the whole realm of material affairs has been remade, and the remaking still goes forward, a gradually unfolding, never-ceasing, modern wonder. It is the miracle of a new world, in which the hum of electric generators and the purr of electric motors chant a ceaseless pæan to men of mighty minds, both of the past and of the present.

Among such men, Charles Proteus Steinmetz stands well to the front. For Steinmetz was a world-builder. That this world of ours is advancing under the magic touch of electricity is due, in several definite ways, to his accomplishments.

He established certain fundamentals that will forever enter into the foundation of this new world—

this world of electrical things. Electricity comes close, to-day, to innumerable folk because Steinmetz worked out those fundamentals, enabling other men to expand the practical applications of electrical energy.

And Steinmetz was an idealist. It was pure idealism that shaped his social philosophy. He sincerely desired a "better world," socially and morally, as well as materially. He never hated his fellow-men; he always loved them and sought to do them good. His life had much of the pathos of the idealist—the pathos of being sometimes misunderstood, and the pathos of sometimes entering the lists on behalf of a cause foredoomed by existing conditions to defeat.

It is Steinmetz the world-builder and Steinmetz the idealist that this volume seeks to portray. It is also of Steinmetz the man of sociable disposition and very human spirit that the author would write; for these things are the most attractive characteristics of his picturesque life and brilliant career.

The author gratefully acknowledges material assistance rendered by Joseph LeRoy Hayden, foster-son of Dr. Steinmetz; Rudolf Eickemeyer, of Yonkers, New York; Charles N. Waldron, editor of the *Union Alumni Monthly* of Union College; Miss Helen Clinton; A. E. Averrett; Dr. Ernst J. Berg; Eskil Berg; Walter S. Emerson, of Yonkers; and the Rev. Ernest Caldecott.

CONTENTS

LIST OF ILLUSTRATIONS

CHARLES PROTEUS STEINMETZ:
A BIOGRAPHY

CHARLES PROTEUS STEINMETZ:

A BIOGRAPHY

CHAPTER I

THE LIGHTING OF THE TORCH

MEN of accomplishment have invariably been the world's thinkers. For to think constructively is to advance, and out of such thought flows the great flood-tide of all human history. From Confucius to Columbus it was thus; and so it has continued, with no sign, even in this age of much turbulence of ideas, that the hoary, yet vitalizing, rule has been invalidated.

Some men overtop all others as men of thought. They become clearly set apart from the men of action, the doers. Their wisdom marches on ahead of the world's civilization, like a great torch leading the way into realms still unexplored.

The sage who contemplates and concentrates is distinct from the executive who performs and directs, although both are essentially thinkers.

Dispassionate observation would seem to justify the conclusion that Charles Proteus Steinmetz belongs in the former of these two groups, among the men who have advanced the world purely by the vigor of their thinking. Indeed, it would appear that he is properly to be regarded as having been a thinker among thinkers. His scientific attainments were the result of intensive study, research, and calculation. He thought deeply, also, upon other questions than those related to his calling of electrical engineer. And, recognized as an authority in the latter, his opinion was heard with readiness in more extraneous matters.

But he was not an executive, or an administrator, in his career. His life-work was not to execute, to direct, to do; it has been to investigate, to study, to think. He represented the thinker, as distinguished from the doer.

Steinmetz was the torch-bearer for electrical engineering in a very definite sense. Not that this great field of human endeavor has had no other guiding Lucifers; but none has lighted the hidden pathway leading to the solution of such deep-rooted problems, which required pure individual brain-work. And Steinmetz had these essentially mental feats to his credit. He untangled complex problems, problems so fundamental that to-day important methods are based upon his work.

Electrical practice and all the colossal benefits to life-battling men and women that have come from electricity have been enabled to grow—to develop and expand, soundly and steadily—because Charles P.

THE LIGHTING OF THE TORCH

Steinmetz lived and devoted his remarkable mental capacities to the service of electrical development. And so he became a man of note; a prominent personage in that realm where precise accuracy is paramount and technical knowledge a prime requisite.

This was not the only thing of remark about Steinmetz. It is just as fascinating to discover that his renown was solidly earned. His rise was the rise that comes from merit.

Steinmetz stood virtually alone in New York City one day in the late eighties, a penniless young immigrant, who could only speak a few words of English. He went from one concern to another looking for employment. When he got it, his pay was less than that of a ditch-digger of to-day.

The only help Steinmetz ever got was the financial assistance of the friend who paid his passage to America in a French immigrant steamer—steerage passage, too, by the way—and who successfully interceded for him at Castle Garden when the immigration officials were on the point of sending him back to Europe.

He was consequently a clear-cut instance of a man who won his way without the aid of influence, patronage, or artificial boosting. Steinmetz got where he did by himself, unaided. But do not suppose that this made him presumptuous. Nor did it overdevelop his ego.

The friendly, courteous manner toward all people, his inferiors as well as his superiors; the kindly, companionable smile that sometimes deepened into a fraternal laugh, a real man's laugh; the thoughtfulness for the good of the human race, which was deep-seated within his heart, left with his intimate friends, and to some extent with his visitors, an impression of democratic simplicity that operated to open the heart of the "other fellow" in friendly response. He was one of those wholly delightful people who somehow contrive to stay wholly human notwithstanding the brilliance of their renown.

Steinmetz, as he finally stood before the world, a master mind among many master minds, a mathematician of tremendous ability, an electrical engineer of deathless accomplishments, was first of all a great scientist. He was other things, also; but his fame rested most securely upon his scientific studies. Hence, in considering the career of this outstanding thinker, it is natural to inquire as to the influences that gave such a personality as this to the world in 1865, the year in which Steinmetz was born of German Protestant parents in circumstances comfortable although by no means exceptional.

His father, Carl Heinrich Steinmetz, had a distinct liking for mechanical matters. He kept himself posted on the scientific developments of his day. He bought books on science, followed new inventions and discoveries, enjoyed studying the lives of notable scientists. He was not particularly mathematical, however, nor was he greatly interested in electricity, which was then but little understood.

He was not attracted by politics or social economy,

subjects which absorbed a good deal of his son's time and meditation.

As for his father before him—the grandfather of Charles P. Steinmetz—there is no record here of any special scientific leanings whatever. The grandfather, Carl Steinmetz, was a country innkeeper in a small town in Poland, although by birth and breeding he was a German.

So little is known of the maternal ancestry of Steinmetz that whether or not any of his mathematical genius was inherited from his mother's side it is impossible to say. His mother was Caroline Neubert, a resident of Breslau, Silesia, at the time of her marriage. She lived only a year after the birth of her son. He never knew his maternal grandparents.

Thus the delicate thread of inherited talents disappears almost immediately into dim obscurity; and we are left to conclude that for the most part the genius of Steinmetz was conferred upon him exclusively at his birth.

In the early years of the nineteenth century, when men still lived who remembered the wars of the Napoleonic era, the grandfather of Charles P. Steinmetz, who was plain Carl Steinmetz, left his home town in the South German province of Silesia and migrated to Poland. He was a young fellow, stalwart and alert, and he was following the oft-repeated practice of going forth to seek his fortune.

Reaching the little Polish town of Ostrowo, he

found need there of an inn. So he stopped at Ostrowo, acquired land, and built himself a house. He
farmed the fields, did something of a business as a
trader, and at length began to prosper. Soon he had
a substantial establishment, of which the inn was the
center. The years passed, and he became wealthy as
wealth was measured in that small place.

Early in his residence at Ostrowo he married a
young Polish woman whose family name was Gawen-
ska. This young woman was to become the grandmother of Charles P. Steinmetz; yet as little is known
of her lineage as is recorded of his own mother. And
it is almost as great a disappointment, for she took
the place of a mother to him through the early years
of his life.

The little German lad who called her "Grossmutter," in later days, as he played around a German
home back in Silesia was to be her greatest trial and
her richest treasure; persistently getting into the odd-
est kinds of childish scrapes, yet unfailingly holding
her affection, for her heart was tender for him.

The family of Steinmetz, the innkeeper of Ostrowo,
was eventually increased by the birth of three sons.
They were August, the oldest; Carl Heinrich, father
of Dr. Steinmetz; and Rudolph. These boys were a
great pride to their parents; and it was a happy, rustic family group that lived there in contentment for
ten years or so after the birth of the youngest son.

Then, in 1846, revolution broke out in the country.
Poland had been partitioned among three European

powers, Germany, Austria, and Russia. Ostrowo was located in that part of Poland which had been given to Germany, although the Russian frontier was not far away. The revolution of 1846 was caused by the rebellion of Poland against all three of the nations that ruled her dismembered territory.

Around Ostrowo the tide of warfare whirled and eddied. In a few merciless weeks everything was changed, even the very aspect of life. The times became exceedingly evil. The contending armies, Polish and German, fought back and forth almost at the threshold of the Steinmetz home. The land was devastated, live stock was carried off, crops were ruined, farm buildings were destroyed. Poverty and despair were left by the battling troops. Pestilence soon appeared, and many neighbors and friends of the family were carried off by disease. Desolation overcame happiness.

The Steinmetz homestead, in the midst of this dreadful maelstrom of misery, was never actually molested. Curious cross-concepts of nationality protected them. When Polish troops were in the region, they guarded the rights of the wife and mother, their countrywoman; and when German troops swept in, they regarded the family as German subjects and so did not disturb them.

But there was almost nothing to eat. Hunger steadily increased, until finally the people had to eat a sort of bread which they made from grass seed.

Somehow or other the Steinmetz family struggled

through this distressing period, until at last the revolution passed. Gradually their former prosperity returned in part. Then the sons began to grow up and to secure employment, which helped the family income. And so, until the boys drifted away, leaving the old homestead, they were again a happy household.

The oldest of the sons, August, served the three years that constituted the prescribed period of military training. At the end of that time he continued in the army for six years longer as an officer in the Pioneer Corps. Nine years of military life entitled him to a pension. The Government also undertook to provide positions in civil life for those who were retired from active military service. Under this arrangement, August Steinmetz got work in Breslau, the capital of Silesia, as a clerk in the government railway office.

The youngest son, Rudolf, became a carpenter. He stayed most of his life in Ostrowo.

The second boy, Carl Heinrich, learned the lithographer's trade. It was a fairly remunerative occupation, and it had just enough of the technical about it to attract him. But work was not very plentiful around the home village, and so he wrote to his brother in Breslau to inquire about a position in that city.

August replied after a while that there was a chance for a lithographer in the railroad office, if Carl could come at once. So the latter bade farewell to his folks, packed his belongings, and made the long journey to the metropolis of Silesia. This was in 1860 or 1861, when Carl Heinrich was in his early twenties and his brother was several years older than himself. It was at a time when Germany was far from being the forward-pushing industrial nation that she became in later years. Instead, the principal occupation was agriculture.

When Carl Heinrich reached Breslau, he found his brother well established there. August had married a young German woman whom he had met after locating in Breslau, a young woman whose name was Caroline Neubert. There were two daughters, little children when Carl Heinrich first saw them.

The family made the new-comer welcome. He took up his residence with them, at their cordial invitation. He found his sister-in-law friendly and attractive. The home life was cheerful, counterbalancing his homesickness.

Then, in 1863, August died from tuberculosis; and in the period of bereavement Carl Heinrich found an opportunity to be of genuine human service to the widow and the two small children. The friendliness between the mother and himself deepened during these days through the bond of sympathy. Gradually a mutual affection budded, grew, and blossomed. And the outcome of it was that at length they were quietly married, about a year after the death of August, Carl Heinrich thus becoming the stepfather to the little girls. They set up housekeeping in an apartment-house in Tauenzienstrasse. And there, on April 9, 1865, was born to them a boy, the third Carl Steinmetz, later to become known to the world and to achieve big things in science as Charles P. Steinmetz.

The name they gave him at his birth seems long and ponderous when presented in full. They christened the baby Carl August Rudolph Steinmetz, with the thoughtful purpose of preserving the memories of his father and his two uncles.

This name failed to stick with him longer than twenty-five years or so. The whimsicalities of university intimates and the Americanizing influences that set in before he was thirty caused the elimination of those two middle names and the substitution of the more cherished "Proteus," conferred in jest, but adopted out of fondness for college associates.

No single prophetic element foreshadowed the remarkable career and the free citizenship which this son of the plain-living, hard-working couple on Tauenzienstrasse was one day to attain in a land across the sea. Nor could they foresee that his native land would not always prove congenial to him as a home.

Yet even at that day the political aspect of Germany was shaping those conditions which were to bring Steinmetz and the realm of the Hohenzollerns to the parting of the ways, the sequel to which was the immigration of the young man to the land of Columbia.

Germany in 1865 was the Germany of Bismarck, Bismarck at his height. A grim, iron-hearted Germany. A Government that was uncompromising in its non-toleration of all counter political thought.

It was a Germany of suppression for liberal thinkers. Hence a Germany of hostility toward the social revolutionists, who were akin to the present-day Socialists in this and other countries. A stern policy prevailed against these idealists; for such they were at the first. Yet, notwithstanding the dark frowns of Bismarck in his Prussian office at Berlin, the social revolutionists continued their forbidden activities in almost every part of Germany.

Four years previous to the birth of Steinmetz, William I, grandfather of the present exiled William Hohenzollern, had ascended the throne as the king of Prussia who later became the first German emperor. And Bismarck the iron-hearted was now solidifying the German Empire in a manner that marked him as a great statesman, even though an intolerant one.

There was autocracy in the Germany of that day; it was not always apparent, but it was always there.

Anyone could perceive that this was an unfavorable national atmosphere for a thinker of the stamp that the new baby on Tauenzienstrasse, Breslau, was to become. For the country was rock-ribbed with political repression. Any sort of thinking that did n't look orthodox to the Government was simply taboo. There was no encouragement for conscientious reflection and untrammeled conclusions.

This, of course, need not necessarily be a hindrance to a scientist; but to a scientist who was also a social economist, politically active, it might become a black thunder-cloud of peril.   And so, in truth, it did.

There are numerous cases of men with exceptional endowments who have been born amid humble environments.   The birthplace of Steinmetz, as it happened, was not so much humble as it was just ordinary, without distinction, without even individuality.

A plain-looking, four-story brick apartment-house, in appearance very similar to a squared-off factory building, was where Steinmetz came into the world. The house looked exactly like scores of others up and down the length of Tauenzienstrasse.   There were rows and rows of them.   A genius might have been born in one as well as another.   It was the dwelling district of families that were well-to-do, but not much more.

Prosaic, workaday reasons led to the selection of this rather unromantic neighborhood, this stronghold of the masses, as the Steinmetz domicile.   Carl Heinrich Steinmetz, the father, wanted to be near his work in the headquarters office of the Ober Schlesische Railroad.   The Tauenzienstrasse region was within convenient walking distance of the railroad offices. Hence, in the economy of every-day arrangements, it was well suited to the needs of the railroad lithographer.

The street was toward the outskirts of the city. Near-by was a pleasant park with a pond and a small stream.   The pond and stream had once been the moat of a fortress, since demolished, and sentries with fixed bayonets had patrolled the ramparts in the days of Napoleon.

When Breslau was besieged by the French, a Prussian general, Tauenzien, had led forth a sortie from this fortress and had driven off the enemy. For this exploit he had been honored by having the street named for him. A street of undistinguished, peacetime apartment-houses, homes of the great, striving middle class!

Only one room in the Steinmetz home looked out upon the cobblestones of the thoroughfare. This was the front room, the parlor, with its three or four large windows and its modestly pretentious furnishings. There were two or three rooms at the rear of the house, each with a window; and a connecting room, or hallway, in the middle, without windows.

In the localized social scale of Breslau, the financial status of the Steinmetzes was to be judged from the size of their apartment. The fact that there were five rooms in it meant that the family was fairly well off. Some households consisted of only two rooms; and there were, of course, graduated levels in between. Also, there were some opulent folk who had risen to the dizzy height of apartments with seven rooms!

This was the outward visible sign of a really fat income in a city where an overwhelming proportion of the inhabitants were apartment-house dwellers. Only the luxuriously wealthy lived in private houses.

Like most people who lived in rented quarters, the

Steinmetz family did not stay in the same house more than just so long. But they were pretty well rooted in the same street. They moved, it is true, after lapses of some years—to another apartment-house in another part of Tauenzienstrasse.

Moving under such conditions was not much of a break. It simply meant the same sort of a house, with the same arrangement of the rooms; for the similarity between the apartment blocks was too great to set them apart from one another.

Sometimes they lived only one flight up from the street, instead of two or three flights. For this, too, was a mark of social and worldly standing. But that was the only difference.

There was no moving about, however, for some time after the arrival of the baby Carl. For one thing, a burden of sorrow fell upon the home when the new infant was only a year old. The mother, in some unexplained manner, fell sick of the cholera, which carried her off with appalling quickness. And so Carl never remembered his mother; his grandmother became the center and soul of the home for him until he began to be a pretty large boy.

Left in his bereavement with a baby son and two small stepdaughters, Carl Heinrich, the father, made the most natural move under the circumstances. He sent word to the old home in Ostrowo, asking his mother, who was now a widow, to come to Breslau as housekeeper for the young family.

She promptly came, accompanied by her daughter,

Julia, Carl Heinrich's sister. And this became the family that the junior Carl—Carl August Rudolf Steinmetz—became acquainted with as he grew older. A busy, yet companionable father, a doting grandmother—"Grossmutter" to him—and his aunt, "Tante Julia"; these were the grown-ups. And there were his two half-sisters, Marie, who was about twelve years older than Carl, and Clara, several years younger than Marie.

Later there was another half-sister, Marguerite, the offspring of his father's second marriage. Thus Carl had two half-sisters on his mother's side and one on his father's side. He had no full-blood brother or sister of his own, however. And, as circumstances would have it, Marie and Clara were no relation whatever to Marguerite, although bound by different ties to the same family.

The coming of the Grossmutter meant a great deal to the baby Carl. It brought into his infant years a kindly, benign personality, to whom he was greatly endeared.

Grossmutter did everything for him that a mother could have done. The plain truth is, she spoiled him. But her love for the child, her "little lad," as she liked to call him, was quite unmeasured.

She had a pet name for him. She called him "Carluszek." It was a term of affection; and by none other was he ever known to her.

The two became the most delightful playmates, comrades who never quarreled, who simply thought the world of each other. To be sure there were occasions when Grossmutter was compelled to assume the rôle of monitor, admonishing the little fellow when, like all small children, he got into mischief. But his childish scrapes, often vexatious enough, almost always occurred when he was playing by himself. When Grossmutter was his playmate, the spirit of joyous and serene good nature ruled their frolics and was the only spirit that prevailed throughout the home. And the child's fancy, coupled with Grossmutter's ingenuity, provided some rare fun for them both.

Such a Grossmutter was a valuable possession, for the speeding days of these first few years represented a tremendous period to Carluszek—the period in which a little child makes its first tentative, delicate acquaintance with matters relating to the great world.

In every child's life the world begins in the home. And Carluszek, thanks to Grossmutter, did not miss, to any serious extent, the best that a home can offer— a motherly personality.

CHAPTER II

AN INNOVATION IN TOY HOUSES

ILLUSTRATING that wholesomely discerning observation, "The thoughts of youth are long, long thoughts," Rudyard Kipling has a short story of India in which a little child's dream of life is the grand motif. "The Story of Muhammad Din" reveals in its brief intensity the ambitions of a baby boy who drew in the dust outlines of great palaces and fine cities.

This is merely substantiation, from a keen student of life, of what all wise folk know to be perfectly true, that children have their own absorbing "world of affairs."

Of such a type of busy, much-occupied little people was Carluszek, his Grossmutter's pet; sometimes, also, her pet problem. There were ideas of an odd, constructive kind knocking around in his childish mind by the time he was four years of age. They suggest, perhaps, some semblance of his absorption later in life in the massive affairs of the world of men.

Most children are fascinated, as was Carluszek, by the idea of erecting whatever their fancy suggests, with building-blocks. Carluszek had some wooden blocks with which he carried out large designs. Most

19

children, also, after building such a toy castle, would have stopped with the carefully reared super-structure.

But this was where Carluszek was different. His active little mind wanted to go further. His toy castle, however grand, was lacking in something that he knew the people in his own house could not do without: it had no lights!

The picture of Carluszek just at this period is assuredly a quaint one. A little child squatting on the parlor floor, surveying his play palace with deep concern because it was doomed apparently to eternal darkness after nightfall!

He gravely informed Grossmutter and Tante Julia that the building was the grand tower of Solomon's Temple, which Grossmutter had described to him out of the Bible. But Solomon would never want to enter at night if he could not see. Would Grossmutter and Tante Julia care to go groping about in the dark? Grossmutter and Tante Julia smilingly assured him they would not, and regretted with him the limitations that haunt the world of toy palaces.

Then they went off and forgot all about it. They supposed it was merely a childish whim that had caught his fancy, and therefore a very short-lived whim, just a "fleeting notion."

But—"the thoughts of youth are long, long thoughts"; and Carluszek pondered over this one until he thought of a way to illuminate the tower of Solomon's Temple!

INNOVATION IN TOY HOUSES

He hurried to the kitchen, where Grossmutter was so busy with her cooking that she scarcely noticed the comings and goings of the little lad. He found a candle and some matches in the cupboard, which he seized and returned with eager haste to his tower of blocks. Down on his knees he dropped, got the candle lighted, and then with patient care guided it in through the doorway of the tower. He pushed it to the very center of the little structure, until the small beams of light shone out from every window opening, like tiny beacons to guide an imaginary king and his royal retinue.

This finishing touch being accomplished, he withdrew his hand and backed off a little, in order to secure the full effect of his handiwork.

Perfect! It was now indeed a finished achievement! King Solomon could look upon it and rejoice; there was ample light for the supposed temple worshipers to gather by night or by day. Carluszek, unconsciously playing the make-believe rôle of a modern illuminating engineer, was fascinated.

As he watched the little spectacle, the wooden blocks began to smoke a bit. But, then, the new lamp smoked, too, if it was turned too high.

The smoke increased until the parlor began to have a hazy atmosphere. And then from the kitchen he suddenly heard hastening footsteps; Grossmutter's voice in amazed expostulations roused him; and Grossmutter herself flew in, rudely wrecking the grand tower and snatching up the candle none too soon to prevent the blackened blocks from actually bursting into flame.

Carluszek received a mild scolding amid the ruins of his demolished tower. It was so exceedingly mild, however, that it did not affect his point of view. He still considered that the necessity of having illumination for a toy tower far outweighed the peril of setting the house afire! Naturally enough, therefore, he built another Solomon's Temple before long, again introduced lights within by means of a candle—and a conflagration was again averted, accompanied by Grossmutter's flustered reprimand, only after the toy blocks had been charred a little blacker.

Grossmutter became perplexed; for her too-gentle rebukes did not in the least lessen the persistency of Carluszek's endeavors. She perceived that it was useless to scold him. All she could do was to exercise vigilance, instead of authority, against the very definite possibility that the roof would be burned right over their heads.

After a time, Carluszek grew bigger and did not continue trying to take light into the lives of make-believe people who frequent palaces built from toy blocks. But at the moment he was the childish prototype of the great torch-bearers who have enlightened the world in all ages of human history. And in this immortal group he himself was to be found within a few brief decades.

The most marvelous object in the Steinmetz household was the new lamp. Carl Heinrich, always fascinated by new inventions, had heard or read about this lamp and had contrived to purchase one. It burned kerosene-oil, producing a light that seemed dazzling in contrast with the dimness of the vegetable-oil lamps then in common use.

That first night with the new lamp glowing on the table was remembered by every one for a long time afterward. The family gathered around in a state bordering on awe as Carl Heinrich adjusted the wick, applied a match, and placed the tall glass chimney in position.

"We must be careful with this lamp," he cautioned them. "The oil burns instantly; and if we are too careless we might cause the lamp to explode."

That heightened the awful delight with which they noted the immense improvement in the illumination—pointing out to each other how much brighter objects in the room appeared, and how much better they could see into the most distant corners. Small Carluszek watched everything with contemplative, inquiring eyes. He watched the lamp as it burned steadily hour after hour. And it was many days before he could see the lamp lighted without giving his whole attention to the proceedings.

But in this overwhelming sense of wonder Carluszek was not much different from the grown-ups in his household, so far as the new lamp was concerned!

Only exceptional occasions justified the use of this lamp for an entire evening, or even part of an evening. The dim illumination of the vegetable-oil burners still served the greater part of the time. But occasionally, when a neighbor or a friend came in, the new lamp was brought out and lighted, to exhibit its brilliancy, the fame of which had traveled around the neighborhood.

So it was, one evening, when a special friend of the family came in, that Grossmutter, who had just been through one of her mildly tempestuous times with Carluszek over the matter of the lighted toy palace, brought forth the lamp.

"It is very dangerous," she solemnly declared, as she placed it on the table, "for it might explode, you know."

They admired it in silence, looking about the room to observe how plainly things could be distinguished. Carluszek also admired the light, watching its unwavering flame for a while before returning to his blocks.

The older folk talked together. Presently the visitor, glancing down at the child, drew attention to the new structure which Carluszek had produced. It was a mill, and it had what Carluszek proudly designated as a mill-wheel.

"It was just an imitation of a mill-wheel which I had put together with two semicircular blocks," said the great scientist, Charles P. Steinmetz, harking back to those childhood days. "It was supported upon other blocks, but it looked round, like a wheel, and my imagination did the rest. At least this mill-wheel was a fine make-believe affair, such as would naturally delight a little boy. Naturally it could not turn—it had no axis—yet I pretended that it would turn if it were placed in a stream. I think I more than half believed that if water flowed over it it would revolve."

Of course, the water was lacking. There was no stream. Therefore the mill and the mill-wheel seemed to Carluszek as incomplete, as unfinished, as had the temple tower without its lights. The mills on the river all had water that made the wheels revolve. That was what his mill needed—water!

"I want a mill-stream! There must be water!" he exclaimed, looking up at the caller in perplexity.

"Very good," said the visitor, jestingly; "here's some water."

And thereupon he offered Carluszek the water-pitcher from the table.

It was done in jest. But before any one could say a word, Carluszek had seized the pitcher and had poured the water over his mill-wheel. For one instant he was delighted with his perfect mill and mill-stream. But the mill-wheel did not revolve; the water merely dripped down upon the floor and flowed over the carpet in a swift, exploring rivulet. Grossmutter, in despair, found her salvaging ability once more called into action because of the little boy's unexpected enterprises.

"Carluszek," she said reprovingly, "I shall have to tell thy father when he comes in."

But it was a mild threat. A brief reprimand was all that Carluszek usually heard of it. Grossmutter never forgot that he was a motherless lad. Grossmutter's heart was big—and Carluszek had the freedom of it.

The bonds of comradeship are extremely strong between the very old and the very young. Indulgent though she was toward Carluszek, Grossmutter was nevertheless a most congenial playfellow with him. And the game that entertained them the most was "going to America."

A small red chest was the vessel. Safely embarked side by side upon this veteran of the waves, the floodgates of imagination were immediately opened, and between them the old woman and the little boy had adventures outrivaling those of Columbus, Magellan, Ponce de Leon, Sir Walter Raleigh, and all the rest.

They forgot they were shut in by the staid walls of the front room on Tauenzienstrasse. They saw beyond those walls, saw the sparkling blue waves of the Atlantic, the deeper, cloud-dressed blue of the sky, the soaring sea-gulls, the gleaming sails of distant ships, and at length a long shore and the smoke of many cities. They had come to America—the golden New World, of which they had heard so much, where so many of their friends had gone from Breslau, seeking a new fortune. Years later, when Steinmetz had in reality reached America, the little red chest was still with him.

Favored of the house of Steinmetz, through the propitious presiding genius of Grossmutter, Carluszek began to show the usual reaction to such conditions. He displayed unmistakable signs of spoiling; and the most infallible of these signs was the outbreak of a violent temper, easily aggravated, but not easily appeased.

His sisters, Marie and Clara, the elder of whom was now about sixteen, suffered considerably from the small tyrannies of Carluszek, who knew himself to be Grossmutter's favorite. He was unwilling that any meal should begin without his presence. He demanded more than his rights in the games they all played together. He played the small usurper upon innumerable occasions. Particularly did he take advantage of his position in the game of collecting pumpkin-seeds, so much so that the situation became quite intolerable.

This game was a much-enjoyed sport that occupied the children of the household whenever a fresh pumpkin was to be cut. The game was confined to a contest to see who could secure the most pumpkin-seeds after the cutting. Usually it was Carluszek, because usually Carluszek raised such an outcry if he were not allowed deliberately to gather in the lion's share that Marie and Clara found themselves obliged to humor him.

In his childish tyranny Carluszek even commanded that no pumpkin should be cut unless he were there to help himself to the greater part of the seeds.

Perceiving that the little fellow had gone too far, even though he was her pet, Grossmutter at length undertook to equalize the distribution of the seeds by having a secret preliminary cutting of the pumpkins, at which Marie and Clara were allowed to secure a quantity of the seeds. Then, a little later, Carluszek was called in, and Grossmutter, having fitted the pumpkin together again, pretended to cut it for the first time by skilfully inserting her knife in the almost invisible crack between the two sections.

One day this trick was exposed. The two halves unfortunately slipped and fell apart before the knife touched the pumpkin. Carluszek instantly understood the deception. There was a perfect storm of tears, and he screamed and wailed for many minutes until his rage had expended its fury in a prolonged fit of sobbing.

Carluszek's sisters were really uncomplaining girls. They seldom grew provoked at his procedure. Perhaps they humored him in their own way as much as did Grossmutter, for they were old enough to understand his childish imperiousness. In their pumpkin-seed game, too, there were usually so many seeds that the two sisters got a pretty good supply, even when Grossmutter did not attempt her bit of trickery, but made the first cutting when all three of the children were there, Carluszek included.

There was another rebellious scene on the day when Carluszek came home to find his sisters already eating dinner, although it had long been understood that no meal should begin until he was there. On this day the two girls were to attend confirmation class and, in order to be punctual, could not wait for the usual meal-time.

Carluszek was fearfully enraged at this disregard of his own "rights." His flood of tearful protests was not even quieted when Grossmutter, helplessly endeavoring to pacify the storm, pretended to have the meal start all over again and to have Carluszek sit down with the others.

Carl had a congenial father; not exactly a comrade, yet an interested observer of his little son's activities and games. He was never too busy to notice what was going on at home. Grossmutter told him about the queer notion Carluszek had of placing lighted candles inside his toy palaces and of pouring the contents of the water-pitcher over his crude toy mill-wheel. Carl Heinrich listened and nodded and watched Carluszek at his play, pondering what these things might signify.

"Doubtless he has a liking for the mechanical, the constructive," said the father to himself. "Perhaps he will follow a craft. Perhaps he will admire progress and invention in the world. We must see how he develops."

Since he was a railroad man, and since boys, from the beginning of time, have always been fascinated by railroad-trains, it was natural that Carl's father should surprise him one day by bringing home a new toy. It was a wonderful locomotive, all equipped with piston and driving-wheels. It used alcohol for fuel and had a really impressive degree of motive power. It did not run on a track, but it would move across the floor, after "getting up steam," with rapidity; its wheels ran right over the carpet! On the tender was painted its name, the Strousberg.

Carl was immensely pleased with this miniature piece of rolling-stock. But his father did more for him. Very cleverly he carved out two small wooden cars, one a freight-car and the other a passenger-coach. He was skilful at this sort of thing, and he was interested to see how Carl would like the transportation idea in his play.

The famous Strousberg was preserved by Dr. Steinmetz after he grew up. He kept it at his home in Schenectady as long as he lived. It would look pretty tame if placed beside one of the magnificent toy electric locomotives, running over a complete miniature railroad system, such as twentieth-century boys possess—if they 're lucky! But it would not need to feel abashed thereby, for assuredly the profession in which its owner distinguished himself has brought to pass that very electrical advancement which has produced, among other benefits for an ever-changing world, the toy electric railroad of to-day!

At times Carl played out of doors, with small friends of his neighborhood. For he should not be thought of as a physical weakling, although unfortunately his body was not perfectly formed like the bodies of other children. Yet he was always healthy and was surprisingly strong. His out-of-doors play was as enjoyable to him as it was to any of the others. And all his life he found great zest in the open air.

The chief playground of the children of his neighborhood was the little park that had been formed out of the old moat. There, in the duck-pond, the boys sailed numerous small boats, or, when the streets were flushed in the summer, they launched these craft in the streaming gutters, with great glee and shouting. In imagination the duck-pond or the gutters typified the broad Atlantic, and the small boats were bound for—America, of course!

None of them thought of any other destination. It was always America. They had heard so much about that great land, knew of so many families where some one had gone off to live there, that they conceived of America as almost the only place any one ever went to besides heaven.

Carl was as happy, as lively, and as imaginative as any of his small chums in sailing his toy boats to that invisible, mystical country which seemed only half real to him and very, very far away—America!

CHAPTER III

TRAVELING TOWARD THE UNIVERSITY

FOR most of us, the great adventures of life began with that memorable morning when they washed our face with painful vigor and brushed our hair with unaccountable earnestness. Our cleanest, dressiest clothes were brought out, although we could not quite perceive the reason for it on a week-day. We began to have an uneasy feeling, a hazy sense of foreboding. We were not supposed to wear our best things to play, yet here we were, arrayed in an unnatural state of finery. And we finally learned that all this was in preparation for our first day at school.

School was a new idea, a new word, heard perhaps indifferently once·or twice before, but allowed.to pass unpondered. Now, however, it immediately stirred our curiosity, doubtless our apprehensions, too, especially if we were deeply attached to the dear home companions. Intuition told us that it meant an introduction to the great world. It foreshadowed a change in our familiar affairs. And we did n't fancy the idea.

Small Carluszek was like that. His days had passed in tranquil, undisturbed enjoyment with those

boon companions, Grossmutter and Tante Julia.
Childlike, he never imagined a time when he would
not see them for hours together. He trustingly supposed that they would always be at hand, and so,
without troubling himself even to raise the question,
he played happily on.

When the inevitable face-washing and hair-
brushing morning arrived at last, he did not sense its
significance sufficiently to ask questions. But Grossmutter revealed, in a single simple statement, what
was in the wind.

"Thy father desires that we take thee to the kindergarten to-day," she explained to him.

Carluszek had no idea what the kindergarten was.
So he did not object when they conducted him to that
strange new place, some distance from his home.
But when he heard Grossmutter say good-by to him
and saw her go away he was seized with dismay. He
had not anticipated anything like that.

The new surroundings and the new kind of play
made him forget for a while the absence of his home
companions. Yet he did not enjoy the morning.
He felt lonely. He missed kindly, indulgent Grossmutter and playful Tante Julia. He was much re-
lieved when they took him home again at noon-time.

But presently he made a dreadful discovery. He
suddenly learned that they expected him to go back
again to the kindergarten in the afternoon. This
was an appalling prospect. It was more than Carluszek could stand; and he gave way to a wild storm

of tears, a despairing appeal, which so touched Gross-mutter's big heart that he was allowed to stay home for another year.

That meant another twelve months of easy, indulgent life, almost wholly devoid of any real discipline. It meant that he would come a little nearer to developing into a mere family pet. In the Steinmetz home the rod was continually spared, and the child was in a fair way to be completely spoiled.

However, a year later, when Carl was five and a half, he was sent to school. His father talked to him once or twice about it, doing his best to make school appear as an interesting place. And when Carl finally did go, he stayed. There was no more rebellion; for school began to appeal to him. It turned out to be just what his father had said—an interesting place.

It was not long before he discovered the fascination of acquiring knowledge. Of course it was not a conscious discovery. But it was sufficient to put a new absorption and a new happiness into his day; and these factors increased as each year passed. While still a young boy his admiration for human achievement was awakened. And this stimulated his taste for reading, so that after a few years he was eagerly examining all the books he could get hold of, especially those dealing with the discoveries of science.

And this fascination of learning, weaving into his life a tingling new interest, was what steered him safely away from the selfish disposition that he would otherwise have developed at the hands of his kindly but too indulgent Grossmutter.

This whole process was quickened by Carl's type of mind. Few boys who ever started to go to school in Breslau had the mental capacity of this lad from Tauenzienstrasse. When his mind began to respond to the first trumpet-call of learning, it responded swiftly. It must have been like the opening of a spring bud when the sun begins to warm its petals. Slowly, steadily, he became aware of the wide world around him, a world crammed with such zest that it made life a more enjoyable experience from year to year.

Thus lifted out of himself gradually, with manifold new interests crowding in, he steadily progressed toward the development of a broad, altruistic character, as a foundation for the building of the patient zeal and keen penetration which were to distinguish him as a scientist.

Carl's elementary schooling lasted three years; and it was elementary in the strict sense of the word. His first acquaintance with arithmetic occurred during this period. It was a memorable meeting, this, of the most horny-headed of the venerable three R's and the boy who was one day to reach the utmost limits of those complex subjects to which arithmetic serves but as a threshold.

For a while it did not look as if Carl was to be particularly proficient at figures. He mastered the first simple problems but slowly; and the multiplication tables, strange as it may seem, gave him a great deal of trouble. He struggled with them for some time before he memorized them. Indeed, so toilsome did his progress seem, so long was it before he could multiply readily, that his teachers began to regard him as a somewhat dull pupil.

Yet he had hardly any more trouble learning to multiply, in reality, than any other boy or girl. He was just at the beginning of his schooling; a small lad, six or seven years old, called upon for the first time to perform concentrated mental work. Naturally the task seemed difficult, even repellent. It required effort of a sort that he had never before been obliged to put forth.

"I did n't like the multiplication tables at first," said Dr. Steinmetz, recollecting this experience. "But it was only because I had to work hard at them, and it was the first time I had ever been obliged to do anything of the kind. I learned the tables mostly by myself; I did not receive much help at home in my school work. They expected me to learn at school and to get assistance from my instructors."

In the regular course of his educational progress, Carl Steinmetz eventually entered the gymnasium, the next higher school. This was a classical gymnasium, which admitted to any of the universities. He was now eight and a half years of age and was registered under his full name, Carl August Rudolf Steinmetz.

TOWARD THE UNIVERSITY

He now went to school from eight in the morning to one in the afternoon. There were special studies, in addition, which absorbed several hours each week.

Here Carl made the acquaintance of various classical subjects. In the first year he took Latin, continuing this throughout the nine years of his gymnasium course. In his second year he added French; and in his third year, Greek.

These languages were thoroughly taught; so thoroughly, indeed, that Dr. Steinmetz to the day of his death knew his Horace and his Homer probably far better than do many college graduates. He could prove this by reciting from memory a passage from either one of these ancient poets.

That is, he could begin to recite a passage. But nobody knows, so it is said, just how much Dr. Steinmetz could recite at a stretch, because nobody in his "audience," as a rule, could keep up with him in recalling the meaning of the fine-sounding syllables. Consequently he was usually requested to desist long before his memory showed any signs of weakening. And this was the fruit of a high-school education acquired half a century before. But, then, Steinmetz was the honor man of the whole gymnasium in scholarship!

Other studies during his gymnasium years included Polish, Hebrew, philosophy, and various mathematical subjects, including algebra and geometry. Polish and Hebrew were not required in his routine course; he took them in his special class-hours in the afternoon. The inevitable soon happened in his attitude toward these latter: mathematics became his favorite out of them all, developing finally into an unfailing fascination, a study which brought him deep delight, withal it meant intensely hard work.

And at the gymnasium he came under Professor Fechner. Fechner was a unique type of instructor, a lecturer on philosophy, dialectics, and critique. An admirer of Kant, his characteristic attitude of mind was to question everything.

Not that he was a downright skeptic. But he declined to accept even well-tested results until he had himself thought the question out and satisfied himself that the reasoning of others was flawless. This, of course, had a touch of presumption about it. Yet it was a policy that stimulated individual research in the highest degree.

Carl Steinmetz, with his eager mind just beginning to open to the stimulus of study, found Fechner questioning even the commonly accepted laws of chemistry and physics. And Fechner, without great effort, imparted this questioning method to Steinmetz. The latter, just then in his early teens, fairly seized upon it as an original and independent mental process that appeared to have commanding possibilities.

And from that time, throughout his life, Steinmetz questioned things, to a greater or less degree. What others accepted he doubted with what seemed utter audacity, yet always for a logical purpose.

For he never stopped with merely questioning.

He never cast doubts just for the fun of doubting, nor assumed an attitude of brazen skepticism and then let the whole matter rest right there.

He questioned in order to find the answer by his own independent efforts. Having asked himself if the established law was indeed the true law, in mathematics, physics, chemistry, or anything else; having inquired within himself if the accepted theory was truly the correct theory, he at once proceeded to make his own independent investigation. Why did he do this? He himself would tell you that it sharpened his mind and accustomed him to go to the bottom of things. In truth, it indicated incessant mental energy, the exhaustless zeal of the real scientist, who must know, unwaveringly, absolutely, from the very ground upward and at first hand all the way, that every step in his reasoning is as solid as bed-rock itself.

Let no one suppose that Steinmetz was wasting time because, in following such a procedure, he may have gone over the same ground that other men had covered. There is evidence enough that much of his life achievement grew out of just this habit of mind.

When he came to investigate the magnetic losses in the iron of an electric motor, he began by questioning the accuracy of data which two other men, Kapp and Ewing, had compiled. Undoubtedly he made many of the same calculations which they had made. But he also pushed on beyond the point where they had left off. And the end of it all was that he discovered the law of hysteresis losses, the most valuable law, just at that moment, which electrical engineering could possibly have secured.

At length Steinmetz reached the end of his gymnasium course. He prepared to graduate, being seventeen years of age, and a young man of diligent habits, with earnest, yet pleasant, friendly eyes. He had already won distinction, for he had received such high marks in his school-work that he could graduate without taking the usual oral examination.

But he did not know this when he began to make his graduation plans. He did not find it out until the school superintendent announced the names of the honor students; and this did not occur until the last moment. Steinmetz, meanwhile, naturally took it for granted that he would be expected to appear at the graduation exercises.

Graduation from a German gymnasium of that day was a grand event. It was as much of an occasion as a high school, or even a college, commencement in America to-day. Relatives and guests assembled in large numbers to see the young men put through their paces. The school superintendent presided, enthroned upon a platform, from which he called out the names of the pupils.

One by one, as their names were announced, the boys walked upon the platform, and, standing before the school superintendent, were subjected to the formal oral examination. It was a considerable ordeal.

They had to "think on their feet"; and think straight, deep, and quickly while their folks and the visitors looked on. Moreover they were strictly required to appear in formal evening-dress.

Steinmetz wanted to look his best for this important function. He anxiously ascertained the rules, particularly as to apparel. Then he prepared accordingly. He went to extreme lengths. He bought a dress-suit!

Even late in life Dr. Steinmetz looked back at that episode with a chuckle and a merry boyish light in his eyes. For, the truth is, he gave a good deal more thought to the matter of clothes, in this one instance, than he did afterward in his whole life.

And he remembers this acquisition of the dress-suit with particular zest. It was the only dress-suit he ever purchased—"and I never wore it!"

Such, indeed, was the case. For when the list of honor students was posted by the school superintendent, the list contained just one name. That name was Carl Steinmetz! He was the head of the class, and therefore was graduated automatically, without further examination. He was not called upon to undergo the oral test. He did not have to march up to the platform, before the audience. And so his dress-suit was never worn, for never again did he feel the urge to give thought to sartorial questions, no matter how imposing the occasion.

And there have been a good many important occasions in his later life. Most men at such times would have desired to cut something of a figure. But not so with Steinmetz. The self-conscious importance of youth, which he felt for a brief interval upon finishing his gymnasium studies, wore off with surprising swiftness within the next year or two. Thenceforth dress was a minor matter with him. Clothes are nothing; the intellect's the thing! Such was his philosophy of pulchritude.

The only definite religious step that Steinmetz took for many years occurred during this period of his life. It was his confirmation in the Lutheran Church, following the customary period of instruction.

This proceeding had absolutely no significance for him. He regarded it purely as a matter of form. Theoretically, the Lutheran Church was the church of the Steinmetz family. But the elder Steinmetz was utterly indifferent to religion, and this had its inevitable effect upon his son. The boy came to regard religious sentiment lightly, giving no thought, after a while, to attendance upon the services of the church. Mature reflection in his older years was to modify this attitude; but as a young man he paid scant heed to the claims of the Christian faith.

It must be said, however, that apparently he received no great inspiration from the teaching preparatory to his confirmation. He afterward recalled distinctly how the clergyman who conducted the confirmation class commented on the place which religion should have in the lives of young men.

TOWARD THE UNIVERSITY

"You may find," said this worthy, "that as you pursue your university studies you will have no use for religion in your own lives. Still, you should not forget that ignorant people need it; therefore every one should respect and preserve religion, since it is necessary to a certain extent."

Of much greater value to young Steinmetz, during these years, was his growing companionship with his father. The two were real chums by the time Carl had began to advance in his high-school course.

The father now held a higher position with the Ober Schlesische Railroad. He had been placed in charge of the preparation of train-orders. The bureau of which he was the head drew up all operating orders issued by the railroad, sending them, when ready, to the lithographing department, to be printed and distributed throughout the system.

The Ober Schlesische Railroad was only one of nearly a dozen lines that converged upon Breslau, which was the railroad center for all that section of Germany. This particular road was part of the through rail route between Berlin and Vienna. It ran from Breslau through Upper Silesia, terminating at a town on the Austrian frontier.

His work with the railroad kept the elder Steinmetz a good deal absorbed. He was at his office from eight to one o'clock, and from three to six o'clock, spending the hour immediately after his noonday meal in resting. He was also on duty, as a rule, during part of Sunday mornings.

Yet he took time to be a companion for Carl. His leisure was fairly devoted to the boy of the household. They talked together in the evenings, and went walking together on Sundays. The father was fond of walking; but he rarely went out alone. Young Carl was almost always with him.

They were mutually interested in each other, and in what each was doing. Carl came to be quite familiar with his father's office. Frequently in the late afternoons he would go down there, and while the elder Steinmetz finished his day's work, Carl would watch from the window as the trains passed and repassed, close by. Then the two would walk home together, perfectly contented in each other's company.

In the summer, instead of going back to the apartment-house, with its hot, shut-in atmosphere, they would walk out into the suburbs a little distance. At one of a number of out-of-door resorts, or cafés, they would sit down at a table in the open air and have a light supper, with beer. In this practice they were typical Germans, for it was a universal pastime to eat the evening meal at one of these beer-gardens during the warm weather.

They talked a great deal on all these occasions. The boy told his father what he was doing at school; he asked questions that occurred to his mind as a result of his studies. And he listened closely to all his father could tell him about the technique of operating a railroad.

TOWARD THE UNIVERSITY

The father shrewdly perceived in what direction the young man's fancy was tending. In his quiet, yet deep, pride in his son, he sought tactfully to cultivate whatever leaning Carl had toward a technical career. Among other things which he did was to give Carl easy access to his collection of books, and this proved an immense delight to the young man.

"My father," said Dr. Steinmetz, "had a number of books, almost all of them dealing with popular science, inventions, discoveries, or natural history. I greatly enjoyed looking them over, and read a great deal in them whenever I had the time. I found among them several books on steamships and steam locomotives. I was greatly interested, I remember, in reading about the building of the steamship *Great Eastern*. This was the largest steam-vessel then in existence.

"After reading of how the ship was built, I searched in periodicals and recent books for an account of the launching. But I was unable to find out anything about this event; so I never knew for sure whether she was successfully launched or not, but many years later I came across a newspaper description of the laying of the Atlantic cable, the article mentioning the *Great Eastern,* which was used as a cable-ship. That was the first I had heard of the vessel since the time when I read about its construction from one of the books in my father's little library in Breslau."

And now the year 1882 was at hand. It brought with it the first sharp change, the first real break in life, which young Steinmetz had experienced. For in this year he entered the University of Breslau, where he was to study for six years. The beginning of university life meant an entirely new set of associates, different surroundings, broader, more serious prospects.

He was obliged to part from a number of friends whom he had made in the gymnasium. Few of the young men with whom he had mingled in the high school went on with him to the university; and of the few who did, all entered other classes than those for which Steinmetz registered. Thus he parted company with those who up to that time had been his schoolmates.

Meanwhile, inevitable changes had also occurred in his home. The kindly Grossmutter had gone away, when he was about six years old, to live with one of her other children. After that she divided her time among her sons and daughters, so that Steinmetz saw her only at intervals. She lived to be seventy-five, dying when Steinmetz was about thirteen.

As housekeeper in her stead, Steinmetz's oldest sister, Marie, then a girl of eighteen, took up the domestic tasks. His next oldest sister, Clara, learned the trade of milliner, doing much of her work at home. His Aunt Julia, who was somewhat younger than his father, had married an organ-builder and had moved to another city.

So it was, as Steinmetz began his university course, that the sense of change, of newness, and naturally of strangeness, was keenly present in his mind. As his was (and always remained) a gregarious nature, a nature which was fond of human fellowship, these feelings were pensive. He could not be indifferent to them. He could not ignore what to him was poignant. For Steinmetz, above all else, was always steadfastly attached to his friends, his surroundings, even the house and the street and the city—the merely material environment—amid which his days were spent.

In this somewhat critical period, immediately following his matriculation at the University of Breslau, his father was his mainstay, the one familiar face which was still there and still the same. The companionship between them deepened through this interval, until it was a very intimate and very fine affection.

From the moment that Carl took up his university work, his father watched his progress attentively. Together they discussed the advanced subjects that Carl was now studying. The young man brought home his notes on mathematical problems, all very neatly written out, with explanatory diagrams, in a purple-hued indelible ink. Some of these notes formed the basis for independent work, which he was doing in an improvised home laboratory, set up in his room.

His father sometimes stepped in, while he was conducting an investigation in this crude workshop, to see what was going forward. Once, indeed, Carl gave his father and the whole family a fire scare. While in the midst of a chemical experiment, the contents of a retort became ignited and flared up suddenly with a muffled report. Carl's father came rushing in, followed by the other members of the household, to find the young student calmly extinguishing the flames and preparing to resume work.

This home laboratory occupied a space, or alcove, beside the entrance to the room. It comprised a stand, for containing vials of chemicals, and a collection of electric batteries and cells. Steinmetz was endeavoring, among other activities, to produce aluminum by electricity, and, in his own words, he "got some fair results."

All the mathematical notes and diagrams, which Carl kept with the exactness of a devotee, finally aroused the admiration of his father. It was obvious to the latter that these loose sheets, constantly increasing in number, were of permanent value to the student. Doubtless, too, he appreciated the intensive work which they represented. At all events he surprised Carl one day by proposing to bind them between hard covers, for permanent preservation.

Knowing considerable of the bookbinder's art, the elder Steinmetz carried out his suggestion, binding the notes neatly and durably into several note-books, which, at first sight, suggest the product of a commercial bindery.

These note-books, with their record of his university work, always remained in the possession of Dr. Steinmetz. They contain about a hundred pages apiece, all written out with marvelous precision in firm, clear characters. They really embody the foundation of that famed system of shorthand which Dr. Steinmetz is credited with having originated in later times.

To him, however, they meant a great deal from a sentimental point of view alone. They constituted, in his eyes, far more than a record of his early studies in mathematics. They embodied most of all a tangible memorial of his father, a treasured keepsake, small though it might be, of the parent who watched him advance steadily in brilliant scholarship with quiet paternal pride and unfailing companionable assistance.

CHAPTER IV

IT would be most inaccurate to regard Steinmetz as simply an electrical engineer. And it would seem still more deplorable to suppose him to have been merely a scientist. He was, in truth, decidedly versatile, although probably not usually so regarded by the public at large, because one aspect of his career, that of the electrical genius, has been unduly emphasized. But his versatility was there. And it developed through a lifetime of opportunity, one door after another opening to him, beginning at an early period. Actually, this started back in his gymnasium years, when he investigated in his own fashion that miscellaneous assemblage of books which he found in his father's scanty but treasured library.

Naturally his entrance into the university was to him like a deep breath of a purer, more exhilarating air than his intellectual lungs had ever encountered before. And it began to stimulate his versatility by first of all rousing his capacity for study to an astonishing degree.

He selected difficult technical subjects and pursued them with the utmost application. In his first year he took mathematics and astronomy, attending every lecture given on both of these subjects. Few students were so intense in their work. It was rare for any one to attend every single lecture in any subject; and under the university system many lectures could be cut with absolute impunity, so far as university discipline was concerned, although the effect might subsequently be disastrous to the defaulter's ambition for a career.

But Steinmetz skipped no lectures. Even though the policy of the institution really favored the drone, he simply did not take advantage of it. He was always in the class-room, always taking notes, always working, even making independent investigations in his room at home. In brief, Steinmetz was the perfect equivalent of what American college men expressively term a grind.

And yet he made merry with the rest of them. The boisterous, care-free student parties that frequently issued from restaurants and cafés to go singing through the streets of Breslau found Steinmetz in the midst; he joined with happy heart in the gaiety of student suppers. He was not given to singing, lacking a musical voice, but that did not lessen his sociable spirit.

He was, and always remained, so completely gregarious, such a lover of the social life, yearning after the society of his fellow-men so deeply, that he was quietly happy when among congenial companions. And he found many such at Breslau.

Moreover, his interest in "the social issue," as he

aptly called it, soon became apparent after he entered the university. In his second year he was studying economics, as well as five other subjects. He had analytical opinions on social economy. And he always loved his classical studies, kept up his classical reading, so that he could converse entertainingly on the old Greek and Roman poets or recite extracts from their works.

He expanded his studies year after year until in his final work at Breslau he was taking higher mathematics, astronomy, theoretical physics, chemistry, medicine, and electrical engineering.

As might be expected, a young fellow of this sort, with a mind so ready to turn to almost any field, so swiftly responsive to every worthy idea, greatly impressed his fellow-students.

The professors were also attracted by this seeker after knowledge. He obviously possessed more than the average amount of zeal, seemed at times to be fairly burning up with it. He asked questions at every opportunity during the lectures. He welcomed the chance for a personal discussion with his instructors.

There were several eminent scholars on the faculty at Breslau, two of whom especially impressed young Steinmetz during his first year, the year in which he attended every lecture in astronomy and mathematics. The lectures in astronomy were given by Professor Galle, renowned as the discoverer of the planet Neptune. An old man in 1882, he was still giving vivid, fascinating lectures, which appealed to the imagination of the young man from Tauenzienstrasse, who ever afterward was a close student of the skies.

The mathematical studies of Steinmetz at Breslau were under Professor Schroeter, an authority on synthetic geometry. It was with Schroeter that Steinmetz performed his first specialized work in mathematics.

When one of the classes in mathematics dwindled down to three members, of whom Steinmetz was one, the latter was quick to seize upon the possibilities in the situation for more individual instruction. The depletion of the class was influenced by the easygoing university rules regarding attendance at lectures. It was also due to the custom among the students of dropping subjects which they found they could not easily keep up.

The professor, in this instance, heightened the individual group atmosphere and put the inquisitive Steinmetz even more at his ease by holding the lectures at his home. There the little circle became very intimate indeed. Steinmetz got into the habit of putting down various questions he meant to ask the professor. And ask he did, until there were moments when the lectures developed into an absorbed dialogue between teacher and pupil. The professor was struck by the keenness of the young man; and Steinmetz somewhat unconsciously reveled in the greater progress he was able to make.

As for the other two students, they found little need of doing more than play the part of audience and fill up their note-books. Steinmetz and the professor, as a rule, completely exhausted the point under discussion!

In his profitable book, "America and the New Epoch," Dr. Steinmetz has brought out the relation of the technical and engineering school to industrial establishments in progressive nations. He has shown that in the case of Germany her technical schools followed naturally the rise of her industrial activity. But in the period during which he was a student at Breslau, Germany had not yet developed industrially. She was still largely an agricultural nation. Her manufactured goods were imported, for the most part, from England.

Consequently engineering schools and technical colleges did not then exist, and higher education was regarded from a very different point of view. The universities, like Breslau, were classical institutions of learning. They had arisen as early as the Middle Ages and were rich in tradition, which reached clear back to those medieval times when modern conditions were undreamed of.

Bearing these things in mind, it becomes easier, perhaps, to conceive of the University of Breslau as it was at that time, from the description of it given by Dr. Steinmetz:

"The university was governed by a rector, who

changed every year. He was always one of the prominent professors, and was elected by the general faculty. The institution was divided into four special faculties—divinity, law, medicine, and philosophy. The fields covered by the first three are obvious.

"Under philosophy came everything that was not included in the first three, from modern and ancient languages to mathematics, science, physics and chemistry, natural history, astronomy, etc. The instruction consisted entirely of lectures; no recitations or examinations.

"When the student entered the university, he went to the office and 'matriculated,' paying a small fee. This gave him the right to hear any lecture given at the university.

"There were two terms, or 'semesters,' per year. At the beginning of the term, a catalogue of lectures was published, giving all the lecture subjects, the number of hours, and the time. The student got a catalogue, looked it over, and decided what lectures he wanted to hear. Then he went to the office, registered for these lectures, and paid for them.

"Usually a large subject was treated by four lectures a week and cost five dollars per term; every professor also gave one or two smaller courses, of one or two hours a week, and these usually were free.

"After the student had chosen his lectures, recorded them in a note-book, and paid the fee, he then, at one of the first three lectures, had the professor autograph the note-book, indicating that the student had attended the lecture. Again, at one of the three last lectures, the book was taken to the professor to have him write his name in it.

"Thus you had to attend the lectures at least twice during the term. Otherwise you did not need to attend; there was no roll-call, but sharply at a quarter past the hour the professor came into the lecture-room, went to the desk, and started lecturing. He continued until the bell rang and the hour was over. Then he finished his sentence and walked out; and so on, through the entire term.

"No one cared whether you attended and learned anything or not. There were no examinations at the end of the term. But after some years—usually at least four—if you wanted to get a diploma, giving you a doctor's title, or enter the service of the Government as a high-school teacher, you had to pass an examination, which covered all the subjects you had attended, or were supposed to have attended. And if you had not attended the lectures, or did not know the subjects—well, that was your misfortune."

There were other respects in which it was all quite different from an American college. The university had no campus, to symbolize its very inmost heart; nor were there any dormitories in which the students, coming from afar, dwelt together, amid the scholastic atmosphere. None of the students came from afar. They all lived in the town with the other citizens, coming daily to the university building, to hear the lectures, and going away again after the lectures were over.

But one great characteristic prevailed at all the German universities, Breslau as well as the others. These were the ubiquitous student societies, which embraced every imaginable subject and numbered virtually the entire student body. Indeed, any student who did not belong to a student society was looked upon as decidedly peculiar.

Steinmetz was not at all peculiar in this sense. He immediately joined a society; and, following his natural inclination, he affiliated himself with the mathematical society, one of the most important of them all.

It was the practice in these societies to confer a new name, or "student name," upon every young man of the university when he became a member. The students only waited long enough to get acquainted with the new member in order to learn something of his personality and habits. Accordingly, after a few weeks, a committee of the mathematical society convened for the all-important purpose of selecting a nickname for Carl Steinmetz, the young fellow who had rather amazed them by starting out to attend every lecture in mathematics and astronomy, who was thoroughly at home with the classics, who had pretty well-formed notions about political economy, and who was withal a jolly good fellow at almost any hour.

Considering these first indications of a versatility which was to increase with the years, it may not seem particularly mysterious to find this committee bestowing upon Steinmetz the student designation of "Proteus," indicating one who changes; for Steinmetz, they noted, could change from one subject to another, from one line of pursuit to some other quite different, with the greatest ease.

No concrete reason for the selection of this name is known; Steinmetz himself thought it was adopted simply because it entered somebody's head at the committee meeting. But it fitted him well, from the students' point of view; and they called him Proteus henceforth.

He always remained Proteus; whenever his name appeared in full, the old student term appeared, too. For when Dr. Steinmetz adopted American customs, he Americanized his name by using the English translation of Carl, and substituting for his two long German middle names the familiar college nickname, with all its memories of jolly times, among jolly companions, in Breslau.

And that is why to-day he is known everywhere that he is known at all—which is virtually throughout the world—as Charles Proteus Steinmetz.

The year 1882 was now well advanced. Steinmetz, at the age of seventeen, had become a "fox" in the mathematical society of the University of Breslau. To be a fox meant simply that he was a first semester man, without rights or privileges in the regularly occurring social frolics. He could take part in the merrymaking, but only in a restricted manner.

The foxes had a separate table assigned them at the restaurant where the society assembled. They were not allowed to mingle with the other students. Over them was placed a fox major, who kept them in order and saw that they obeyed the rules.

If there was a dull moment in the swing of events, the president would call upon the fox major to entertain. Then the foxes were lined up and required to sing a song, the other students listening in a critical mood. This was always a harrowing moment for young Steinmetz; for he never could sing, either then or afterward. Of course he tried, as did all the foxes. And naturally the melody was usually mangled beyond recognition, while the upper semester students howled their derisive comments. Finally, if the rendition was utterly offensive to artistic tastes—which was frequently the case!—the foxes were all penalized. They were "sent into the beer," as the saying went.

Literally that meant merely that the offender was obliged to drink his beer and keep still for a specified time. Every member had his glass of beer, for this was the universal custom. In the student societies, however, it was drunk in strict moderation, even hedged about by formal regulations. To be "sent into the beer," that is, ordered to do nothing but drink in silence, was a disgrace. The German student idea of drinking was to drink socially, with conversation, repartee, and care-free jest or song. To drink alone, or to drink in silence, seemed churlish and was seldom done willingly.

Steinmetz found that, as a fox, he was held down pretty severely. Yet he also found that although he had no rights he could adopt an "old man," one of the older members, who would thereby become the protector of the fox that adopted him. If any fox were abused, by being sent into the beer too often, the fox's tormentor might meet with swift retaliation at the hands of the "old man" of that fox, who could send him into the beer in turn.

All this was taken by Steinmetz in the spirit of a good sport. In truth, he thoroughly enjoyed the meetings of the society, which mingled, in the range of a single evening, a business session, a scientific discussion, and several degrees of merrymaking periods. He thus describes these student organizations:

"There are some varied misconceptions of the German university life of those days, created by humorists, like Mark Twain, who never intended to describe the university life as it really is, but rather the funny and startling side of it. Thus they have given the impression that the students all run around with colored caps and belong to student societies which devoted themselves principally to fighting duels and drinking beer.

"Such societies as these did exist, but they represented an insignificant minority of the students.

They had practically no effect on the university life as a whole. To be sure, in some of the smaller western universities they were a little more numerous, but even there they were a minority."

A small, plain room in a restaurant near the university was the congenial rendezvous for the mathematical society. There, one night each week, they assembled, around 8:15 o'clock. The president formally opened the business meeting, which lasted for the first hour or so. There were usually various business matters—the election of new members, the purchase of books for the library, or the selection of a date for the annual kommers, or festival.

About nine o'clock the scientific section of the program opened. By that time some of the alumni members had arrived, and occasionally one or two of the professors came in to hear the scientific discussion.

This consisted of a paper, prepared and read by one of the students. It might be a review of some newly published mathematical work, or the presentation of a new investigation. Then would follow a formal discussion, after which some smaller mathematical problems were propounded and the solutions worked out.

Between ten and eleven o'clock, the more serious phase of the evening concluded. From then on it was a time of relaxation, becoming less restrained as each loosely defined period of the festivities passed.

The scientific session was superseded by the *commercium,* or social meeting; this was displaced by the *fidelitas;* and from that the revelry was carried on into the *suitas.*

There were virtually no restrictions by that time as to the character of songs or stories. The tinkle of beer-glasses was more insistent, too. During the business and scientific sessions it was not considered good form to drink much beer; but throughout the merely merry moments, beer was inevitably called for. There were even exaggerated student formalities, associated with the beverage, such as a "beer president," sometimes a "beer court," and constant ceremonial beer drinking by the students to each other, or by the society as a whole to the mathematical society of some other university, to which notice was sent that the Breslau men had "drunk a half" (half a glass) in its honor.

So through the latter part of the night and into the gray of the morning it was pure fun and frolic. Yet the seniority of the students was preserved throughout, although by midnight the party had dwindled considerably. Alumni had departed; the older students had gone to a large extent; those who remained were the young, gay roisterers who were out to "make a night of it," with utterly care-free abandonment to joviality. Dr. Steinmetz recalled these student larks with a zest and a detail that leads one to suspect he more than once joined in making a night of it himself.

BRESLAU STUDENT REVELRIES

"The *fidelitas* continued until about one o'clock,"
he says. "Then it ended, and most of the fun-
makers went home, but a few remained while the
'*suitas*' was opened, so-called because there were no
more rules, and no limitations to the kind of songs
and stories.

"About two o'clock the last of us left. But that
did not necessarily mean going home. There were
a number of all-night restaurants, known as Vienna
cafés, which started business only after eleven
o'clock, when the theaters closed. Our favorite
Vienna café was only a few blocks from the regular
meeting-place of our society. Eight or ten of us, all
very jolly and chummy, one of the party included
being a particular friend of mine, a student named
Henry Lux, and called 'Hinz' for short, liked to go
there to drink coffee, mixed with rum, and have an
appetizing lobster salad.

"It might seem that this was extravagant for penu-
rious students. However, it was not as severe on
the student purse as it sounds. The café proprietor
served a very fine salad, but made the portions extremely small.

"At the place where the Vienna café stood, two
night-watchmen met, at the ends of their beats. We
were on friendly terms with these officers. If we
met them we always invited them to accompany us
to the café. Then we would send out a cup of coffee
to the policemen.

"We kept on with our jollification by visiting,

about three or four o'clock in the morning, a most congenial place which we knew only as 'Mr. Lichtenheim's.' We called it that because Lichtenheim beer, named after the city in western Germany where it was produced, was served there. As we did not know the name of the genial proprietor, Lux suggested calling him Mr. Lichtenheim; and so Mr. Lichtenheim it was.

"There we could get certain delicacies which were not to be found elsewhere. These included a delightful variety of mussel, or fresh-water crab.

"The crabs were served to us in a big casserole, after some generous-hearted member of the party had offered to stand the bill. When the casserole apeared on the table, everybody made a grand dive for the crabs. Each seized all he could lay hands on. When the scramble was over, perhaps the student who was paying for the feast had got hardly any crabs at all. Sometimes I offered to be the good fellow—with this result!

"At this place we thought nothing of borrowing from the waiter, or from Mr. Lichtenheim himself, to pay for what we ordered. We were very intimate with our regular waiter, who often had more ready money than we ourselves. But when this man was off duty, and a strange waiter served us, we had to borrow from Mr. Lichtenheim—who never refused us. He understood university students.

"It was usually about five o'clock in the morning when we finished at Mr. Lichtenheim's. But nobody thought of going home then. 'Keep on and see the night out,' was the watchword. So we would tramp into the suburbs, a little distance up the river, rouse a sleepy innkeeper, and demand breakfast. We could get a steaming cup of coffee there, if nothing more."

That was the last scene of their all-night hilarity. With the coming of daylight they hastened back to the city and the university, arriving just in time for the first lecture at eight o'clock. On Saturdays this was usually the only lecture scheduled; and despite their overtaxed stomachs and sleepy brains, the students managed to stand it. Steinmetz in particular kept himself in better mental condition, for he never dissipated in the real sense of the word.

No sooner was this lecture over than the students of the reveling party hurried out to the nearest restaurant for their "second breakfast"—and then to their lodgings for a good sleep.

CHAPTER V

IN six years of diligent activity at the University of Breslau, young Carl Steinmetz emerged from the quiet grind of his first semester to become the center of a risky, determined student political movement, which ultimately cost him his university degree and came near to landing him in prison. The contrast in this university career of a student so brilliant that the authorities of the institution dared to make covert efforts to shield him from the wrath of the Government is not only pronounced; it is fairly startling.

Steinmetz himself looked back upon it with full appreciation of all that his decisions and deeds involved, and all that they cost him. "That period," he remarked more than once, "was perhaps the most exciting of my whole life."

Allowing no interruptions to his university work, studying with an intensity that awed his fellow-students and astonished his instructors, Steinmetz nevertheless developed in other directions that had nothing to do with scholarship attainments. As a result, we find him classified far differently at the end of those six years than he was at their beginning.

A STUDENT SOCIALIST

From a quiet, inoffensive fox, obediently following the dictates of the fox major in the mathematical society, he emerged as a fervent, uncompromising student Socialist, maintaining his allegience to the movement secretly for nearly four years, but suspected and finally apprehended by the police, to whom he appeared as a dangerous ringleader of young, anti-government fanatics.

A fanatic he may have seemed, for he was in the process of acquiring intense convictions on certain great human questions.

But his fanaticism did not become an excuse for slighting his studies. First and foremost he was a scientist in the making, bent, at the moment, upon diligently developing his mind by the systematic acquirement of knowledge.

This is significant for several reasons. Above all else it seems momentous because it is the spirit that led Steinmetz first to interest himself in the great subject of electricity and electrical engineering. That occurred even after he had begun to espouse the cause of the social revolutionists and had thereby engaged in much additional work, aside from his zeal from university accomplishment. He did not know it at the time, but the subject of electrical science was in general to concern him more closely throughout his life than any other, not excepting the Socialist creed which he ever afterward upheld.

Steinmetz first took up this study at Breslau in the spring of 1886, while he was in his fourth year at

the university. It was two years after he first affiliated himself with the German Socialists. At this
time he was zealously attending the secret meetings
of the student Socialist group and reading and cir-
culating Socialist literature, in direct violation of the
Government's manifesto.

Possibly if Steinmetz had gone to any other university than Breslau, which was in his native city, he
might not have taken up electricity as early, or as
thoroughly as he did. Such was the insignificance of
this subject among German educational institutions.
None of them had a course in electrical engineering.
Indeed, we are led to conclude that no other university in all Germany dealt in the slightest degree with
this tremendous, but still virgin, field.

At Breslau, electricity was included in the study
of physics. But it was not what might be called a
popular course. Very few students attended the
lectures.

And yet there were characteristics about it which
vividly appealed to young Steinmetz, with his love
for mathematics. Moreover, what little had been
done in applied electrical science immediately aroused
a strain of scientific speculation in his fertile mind.
To him these things foreshadowed in some degree the
effect that such a manifestation of boundless energy
would surely have upon mankind in general. This,
as may be supposed, was enough thoroughly to fas-
cinate a young fellow like Steinmetz; so, although

he never dreamed at the moment that it meant a career for him, he began to study everything there was to study about electrical engineering.

The world at large was just then beginning to conclude that electricity was indeed something more than a freak plaything of the laboratory. Seven years previously Thomas A. Edison had brought out his incandescent electric lamp, thereby making possible the first widespread commercial use of electric current. And Nikola Tesla, young man of twenty-nine, was entering upon a period of intensive research into methods of producing electric currents and into their characteristics.

Yet everything regarding the practical use of electricity was in a hazy, tentative state. Edison's first lamps were crude; equally academic was his first electric light system. And neither Tesla nor any one else up to then had devised such a thing as an alternating current motor.

Steinmetz took up electrical engineering, as it were, with an open mind. He was willing to learn anything new in applied mathematics. And that is exactly what this subject held out to him.

But he certainly had no idea whither it would lead; he did not then have the slightest intention of becoming an electrical engineer. As far as his life ambitions had shaped themselves at all they had led him to drift toward his one real fascination, mathematics. He was thinking, at this period, that he would prob-

ably like to teach mathematics. Perhaps his big aspiration was to reach the height of a professor of
mathematics at one of the universities.

Picturing his life goal thus, he began to study the
great mystery of science, the unknown thing—as it
then was—called electricity. And his first original
scientific investigation was in this general field.

It occurred during the next winter, when, with
several other students, he sought to determine the
constant of the terrestrial magnetism at Breslau.
They hit upon a bitterly cold day for their work; and
as the ironwork in the university building interfered
with the magnetic needle, they sallied out, perforce,
upon the frozen river, led by the intrepid Steinmetz,
who considered a frost-nipped nose or ear only a gentle form of martyrdom for the cause!

For a while they made progress. Steinmetz did
the greater part of the work, taking all the records
of the observations. The warmth of his zeal enabled
him, apparently, to withstand the cold better than his
companions, for at length he found himself alone
upon the ice. His helpers had departed to a near-by
café to get warm, leaving the instruments with
Proteus.

The latter was so engrossed in his investigation
that he let them go and continued to work on alone,
ignoring the extreme cold. Finally, however, their
prolonged absence aroused his suspicions. He went
in search of the deserters and discovered them snugly

ensconced in a warm room in the café, drinking beer and playing cards. They had not the slightest thought of returning to the freezing quest of the terrestrial constant.

Amid much merriment they informed Steinmetz of this decision. He raised no objections; and thereupon the instruments were packed away and the expedition dissolved itself into a typical student party. The climax came a day or two later with the discovery that the small amount of data they had obtained was entirely erroneous on account of the steel rim on Steinmetz's eye-glasses, which had disturbed the magnetic needle!

Among the chums whom Steinmetz acquired through the mathematical society, Henry Lux, who was known as Hinz, has been mentioned. This young man was about a year older than Steinmetz; a thoughtful, inquiring, intellectual fellow, hailing from Upper Silesia. He brought along with him most of the daring ideas of a typical political socialist of those times.

Lux was not the only student at Breslau who had accepted these principles. Consequently, as was to be expected, he sooner or later met one or two others who thought as he did, and this group formed itself into a loosely organized club. It had no written constitution or declaration of purpose. It met without ostentation, even camouflaging its sessions as social gatherings. In short, it was, in effect, at least, a student Socialist society, well knit together by a common spirit, keenly felt.

Before long the group gained several adherents among the residents of Breslau, men who were actual working Socialists, dyed in the wool, and affiliated with the national Socialist party. One of these was a physician. Another was a bricklayer of an unusual order of intelligence. He devoted most of his spare time to editing the Breslau Socialist newspaper, known as "The People's Voice."

Steinmetz was first invited to one of these Socialist affairs by Lux a year or so after the two had made each other's acquaintance in the meetings of the mathematical society and the revelries incident to that organization.

"He did not try to win me over to the Socialist way of thinking," says Dr. Steinmetz. "He merely invited me to one of the meetings, informing me that it was a free-thinking group and that he thought I would be interested. So I went with him to my first Socialist meeting, to listen and afterward to meditate upon what I had listened to."

In these tea-drinking circles—for the group met at various homes and invariably indulged in tea and cake, because of the paucity of funds—the young mathematical student absorbed the doctrine of German Socialism. It was not Socialism as generally understood in America to-day. It was an idealistic creed, rosy-tinted to the mental vision, aspiring unto perfection for human society, and fired with all the enthusiasm of youth.

And this just fitted the principles that Steinmetz had quietly cultivated, more or less earnestly. For he had a strong utopian leaning; and these young men were really utopians, especially at the beginning.

It seemed to them—and it seemed likewise to Steinmetz—that through the Socialist movement they could remake the world into something better than it had ever been before; could root out the sore spots in human society; could put an end to injustice; could teach men practical, daily good will toward one another. At least they were sincere, which also appealed to the young man from Tauenzienstrasse.

Behold here, then, the Socialism of Steinmetz, not only during his university days, but ever after! Notwithstanding his affiliation at the close of his life with the American Socialist party, which is frequently assailed as the expression of selfish class interests, Steinmetz was not a believer in nor an exponent of selfishness, certainly not in human relations. He remained an idealist, as he was when he sat around drinking tea with the Breslau student Socialists and listening to doctrines that solidified the tentative, uncertain fancies of his own mind into earnest convictions. He wanted to see happiness universal. He believed that Socialism in its pure state is sufficiently altruistic to bring this about. And he made sacrifices for his principles.

Sacrifices inevitably came the way of those who ag-

gressively espoused this cause in the Germany of
the eighties. It was impossible to be an active Socialist without running athwart the Government at
Berlin, with its severe Bismarckian policies. Germany was at peace with all other countries, yet Socialism was persecuted as something abhorrent to the
Government and intolerable for the nation.

Like all causes in similar circumstances, Socialism
flourished under repression. And the repression was
especially severe in its local application, because Bismarck was endeavoring to nip the obnoxious movement in the very bud.

There was a cold-blooded reason for this. The
early beginnings of the organization, as initiated in
its local branches, was its most successful point of
attack. It was the stage at which the Government
could move with the least embarrassment to itself.
There was then more freedom of action, for an
autocratic intolerant régime, than was the case when
the Socialists became powerful enough to elect deputies to the Reichstag.

"Representative government in Germany was so
strongly intrenched," Dr. Steinmetz explained, in
discussing the subject, "that it was impossible to eject
a deputy from the Reichstag once he had been duly
elected by his constituents. It made no difference
what his political principles might be. If it was the
wish of his constituents that he should sit in the Reichstag, that was sufficient.

"Thus Germany really had a more truly represen-

tative government than we have in America to-day;
for the electors could express their wishes by their
votes, and those wishes were supreme. The different parties therefore had merely to increase their
numbers, by education and solicitation, to the point
where they could control the elections, and they were
then assured of representation in the Reichstag.

"It proved to be so in the case of the Socialists,
who were rapidly becoming very numerous throughout the land—made so by the severe laws passed by
the government in an effort to put the entire movement out of existence."

Steinmetz had now entered upon a policy of action
which was eventually to terminate his placid pursuit
of scientific knowledge and bring his life into a more
turbulent period. It gradually developed in him a
secret watchfulness against possible apprehension.
The conclusion of this period came in a furtive flight
from home and country; for flight is the only term
for his precipitate departure from Germany four
years later.

Having cast in his lot with the student Socialist
group after hearing their views and reflecting upon
them, Steinmetz rapidly acquainted himself with the
scope of the social revolutionary movement. He
found the student Socialists numbering not more
than a dozen; among them being two young medical
students, then well along in their studies, and one or
two scientific students, as well as the Breslau physi-

cian and the bricklayer-editor, already mentioned. There was also an ardent young supporter of means, the son of a wealthy banker.

A year or two previously, two rather famous student Socialists of Breslau, the Hauptmann brothers, Carl and Gerhart, had been members of the organization. Steinmetz was to meet these two later at Zurich.

The enthusiasm of Lux dominated the little club that gathered just so often to debate upon the best means of advancing the cause. There was abundant zeal; but everything was carried on under cover.

There was always a sense of anxious watchfulness in the air when they came together. However careful their plans for a meeting, however much their real purpose might be screened, however cleverly the word might be passed around, there was always the lurking possibility of discovery by the police. Any unfamiliar sound during their session would put every one on edge, fearing the worst for a tense moment or two. A quiet step outside the door, a movement of the latch—and instantly every head jerked around in a half-panic to see whether it was friend or foe approaching.

"This habit of looking around quickly when the door is opened and some one enters the room in which I happen to be, still persists to some extent," Dr. Steinmetz says, in speaking of these experiences. "For a good many years I would instinctively feel a spasm of the old dread. Even in comparatively recent days I have caught myself glancing up with unnecessary suddenness when I heard somebody entering the room. No one has noticed this trick, since I know that no one in America has suspected that I had any such feelings; and I have always recovered myself before doing anything that might appear strange. But we carried that constant suspicion in our minds for so long a time that it was not easy to shake it off when I found myself no longer liable to a surprise visit from the police."

So frequently did the announcement of a meeting of the group come in the form of a simple invitation to a tea-party that the young Socialists might have been known as the tea club, in lieu of a more distinguishing title. They came together as a few socially inclined souls, desirous of a chat and a bit of good cheer. But the serious purpose behind their gatherings was always felt. The secrecy was in reality to their liking; it made these young fellows feel the adventure of the undertaking to which they had committed themselves. Actually it would have been extremely difficult to obtain any real evidence against them.

Here, in this secret little circle of university youths, blazed the fires of intense altruistic zeal. Each of these young men had dedicated himself to the services of society. They sat and talked and sipped tea with earnest-eyed, eager-eared belief in their own and each other's theories as to the earthly salvation of the human race.

The soul of Steinmetz was as a slumbering volcano, needing only a spark to set it off, in the presence of such a group. And here he found the spark. It is little wonder, therefore, that the social revolutionary movement—or the social evolutionary movement which, he observed, it later became—should be to him as a religion.

Occasionally there was a burst of pure boyish hilarity in the midst of their intensive deliberations. On one occasion, when they were gathered in the home of one of the group, they heard some one coming up the stairs of the apartment-house and approaching the door of their meeting-room. Then the latch rattled and the knob turned.

They all jumped sharply and turned around with nervous haste, thinking that the police must have found them out at last. But the next moment their dread gave way to mirth, for in dashed one of their own number, a late arrival, his face aglow with anticipation, holding out a large bag with a certain bake-shop fragrance about it.

"Doughnuts!" he exclaimed. "Doughnuts for everybody! We 'll have them with our tea!"

There was a chorus of approval. The doughtnuts were passed around. Everybody had some. There were plenty, too, enough for a second serving.

At the height of the unexpected feast the provider of this bounty explained that a kind shopkeeper had given him the doughnuts at a reduced price, because they had been in the shop for a couple of days, and were "just a very little stale!"

Steinmetz rather emphasizes the utopian leanings of nearly all the men in this group, particularly the students. It indicates that he himself found in this the biggest fascination of the entire movement.

These young men would have mended all the bruises of this poor old world at one wave of a magic wand, had they possessed that benevolent power.

"It was not until later," Dr. Steinmetz once said, "that these young men became political Socialists. But when we did go in for that sort of social development, then we put ourselves in all the greater danger from the Government.

"Just at that time the Nihilist movement in Russia was at its height. It had caused the assassination of the czar in 1881. Poland had become deeply tinged with this teaching, and much Polish propaganda had filtered through into Germany.

"Consequently the German Government was exceedingly watchful of the Socialists, among whom it was feared that this radical theory might take root. Severe measures were adopted against the appearance of any Polish activity among the German people.

"One of our group at Breslau was a Pole. We heard more or less about Nihilism from him, but to our more orderly way of thinking Nihilism did not appeal. We did not adopt Nihilist principles, or follow those methods."

One evening, years afterward, when Dr. Steinmetz had become well known as the electrical genius of Schenectady, we were talking at his home about the social revolutionary movement in the Germany of the eighties.

He had been recounting those experiences and his part in them; revealing, withal, the uncompromising fight that was going on between Bismarck and the Socialists, which resulted, Dr. Steinmetz observed, in the defeat and retirement of Bismarck.

"How is it," he was asked, "that this spirit of antagonism did not persist through the recent war? How is it that the German Socialists, once unalterably hostile to the Government, so thoroughly supported it as to result in a united front by all the German parties during the war against foreign enemies?"

"It was due to the concessions made by the kaiser to the social revolutionists," he replied quickly. "The kaiser conciliated the masses. Thus Socialism was changed from a revolutionary to an evolutionary party. The Government made such concessions as insurance against old age, sickness, and unemployment—the three great fears of the masses of the people. Thus the kaiser attached the social revolutionists to the cause of the monarchy, and by this liberal attitude he so won over the Social Democratic party of Germany that when the war came they supported the Government as strongly as did any other party."

It was in the period immediately previous to this governmental change of front, when Bismarck was still in power, that Steinmetz, as an enthusiastic young student, threw in his lot with the hounded Socialists of the old Germany. In doing so he witnessed, and in his own field of opportunity assisted in, the building of the Social Democratic party which wrought potent industrial and economic changes in the nation, as he has outlined in his "America and the New Epoch."

And in doing so he utterly changed the course of his own life by the train of events described in the succeeding chapters of this volume.

CHAPTER VI

HOUNDED BY BISMARCK'S POLICE

AS merely idealists, practical only when they they had to be, the young men of Breslau who first took up German Socialism did not arouse the hostility of the Government. Bismarck let them alone as long as they simply dreamed of, and talked about, a better world, a more perfect human society. It was when they went further than dreams that the clash came.

The dreamers had already awakened sufficiently to adopt a practical program—which meant, in effect, that they had embarked upon political activity— when Steinmetz became one of the group. They had already allied themselves with the Social Democratic party of Germany. Hence came the essential secrecy, for Bismarck's agents, the police, were watchful for evidence of seditious plottings.

The old idealism had not been wholly surrendered by any means. In the first few months of Steinmetz's association with the group it frequently cropped out in conversation and discussion by reference to a utopian Socialist colony in California. This experiment in practical idealism, or idealism applied

to practical affairs, was geographically known as
Icaria.

Steinmetz heard his fellows among the student Socialists relate the story of Icaria, as they understood
it. He was told how the student Socialist club had
despatched a delegate to Icaria to investigate the colony and report as to its success. The expenses of
this undertaking were so large, from the point of view
of the struggling students, that it had to be deferred
for a while. The banker's son was the chief financial
bulwark in the matter, although all the members contributed as much as they were able.

Thus, for a while, Icaria was to them the embodiment of all that was most desirable in earthly life.
In the light of the rather meager information they
had at hand, Icaria was actually Utopia. They were
dreamers, and here seemed to be the rosy fulfilment
of their dreams.

But the inexorable awakening came when the delegate returned, bringing the disheartening news that
the colony had proved impracticable; the experiment
had been given up and the settlement abandoned.

"It turned out to be a failure," said Dr. Steinmetz,
"just as any Socialistic community in a capitalistic
society must inevitably have been a failure. So this
report of our delegate was a disappointment to the
student Socialist club. But for some time after I
joined them, there was discussion concerning Icaria
and the reasons why it had not succeeded."

An incidental effect of this experience was to set

many minds to thinking about America. There was a rosy haze of possibility clustering about this name; moreover, it had a known element of permanency. From all reports, America was a land favorable to ideals, at least a land congenial to individual ambition. Icaria, they realized now, was an illusion; but America was a reality.

This thought, transitory at the time, nevertheless lingered in the mind of young Proteus, adding one more unconscious impulse to consider America as the most plausible destination of those who chanced to travel abroad. Naturally he did not suspect what the future contained in this respect for him personally.

It was the affair of the Lassalle photograph that first provided the Breslau police with a definite clue against the student Socialist group. This incident occurred in the spring of 1884, not long after Steinmetz had cast in his lot with the group, enthusiastic over the discovery that here were men who thought, in the main, as he did upon the condition of society.

It has been mentioned that the student Socialists included two medical students. One of these was graduated that year and planned to leave the city upon completing his studies. In his honor the society held a farewell conclave, surrounded with the usual secrecy, although, in one respect, as will be seen, the required caution was not strictly main-

The famous photograph of the student Socialists of Breslau grouped around the bust of Ferdinand Lassalle. This picture was confiscated by the government agents. Steinmetz is standing at the right

tained. Since this was one of the few departures from among the members of the circle, it was proposed that the entire group should be photographed; and to insure a lasting token of their association together in the one cause, which interested them all, they grouped themselves around a bust of Ferdinand Lassalle, the noted German publicist and Socialist.

In this manner the photograph was made, and enough prints were ordered to provide one for each member. These pictures were greatly valued by the young men, since they furnished common keepsakes as well as a reminder of the principles to which they all had subscribed.

The photographer, in his professional zeal, asked permission to exhibit the photograph in the window of his studio. And right here is where the student Socialists, with all their watchfulness, were guilty of an error in judgment: they granted the requested permission. The telltale picture was displayed in full public view. And to any one familiar with German political events Lassalle was easily recognized as a sponsor of Socialism, which was abhorrent to Bismarck. Yet here, in plain sight of all who passed, was a photographed group of young Breslau students stanchly surrounding the Lassalle bust and thereby obviously proclaiming that they had made the Lassalle cause their own.

Of course it was too much to expect that such a self-evident banding together against the policies of the Imperial German Government would escape police vigilance. From all accounts it was n't noticed immediately by the authorities, who were even then trying to make an investigation of social revolutionary activities among the students of the University of Breslau, without, however, making any material progress. When the condemnatory photograph was eventually observed by the police it was recognized by them as "important evidence." As a matter of fact it provided definite information against the inner circle of the student group—the first link in proceedings that finally led to the apprehension of the most active individuals, several of whom, including the indefatigable Lux, were under arrest within two years.

As for the fatal photograph itself, the police industriously rooted out and confiscated all the copies which they could discover. They endeavored to locate and identify every man who appeared in the group. No prints of the audacious picture must remain in existence! The photographer must destroy the negative! And in truth so energetic did they become that only one or two prints escaped the dragnet.

Steinmetz managed, by much secrecy and judicious discretion, to preserve his own copy. He was not questioned as were the others in the group, for Steinmetz had not yet attracted the attention of the police as a "dangerous agitator." Consequently he had a sly opportunity to secret the picture; and he always retained it in his possession—one of the reminders of "the most exciting period of my life."

Far from feeling timidity toward the ever-prying police, the audacious student Socialists, and indeed all the Socialists of Breslau, did not hesitate now and then to play a joke on their stealthy opponents. As a rule, all activities proceeded under cover—not through fear of the authorities, but to avoid that interference which would have hindered and set back the progress of the party. On one of the rare occasions when the anti-government agitators came out into the open, Steinmetz took part in a merry little game of hide-and-seek between the Socialists and the police. He recalled it thus:

"During the first winter that I belonged to the student Socialist club I attended nearly every meeting. In the following summer the group became strongly affiliated with the German Social Democratic party. About this time I made the acquaintance of the Breslau Socialist who was editor of the local Socialist paper, 'The People's Voice.' I also met the editor of a popular science magazine who had come to Breslau to live.

"That summer I attended several large gatherings of Socialists in Breslau. Sometimes these were honored by the presence of one of the Socialist deputies in the Reichstag. These were always unusually important occasions. Every one interested in the Socialistic cause was invited. There was usually something of an audience.

"Naturally, when so many were interested and invited, the police became aware that something was going to take place. It was not very difficult for them to find out that on a certain date a Socialist deputy would visit Breslau; and they knew, of course, that this would mean a meeting at which the deputy would speak. The police were expected to prevent such affairs—if possible.

"When the day arrived, those invited to meet the deputy would gather rather openly, and the entire party would set off for a stroll into the country. Police officers who appeared to forbid a Socialist meeting would follow along. The Socialists and their friends represented themselves as out for an excursion. They would really carry out this supposition by finally stopping at some country tavern. Everybody would go in and order beer. The police would do the same.

"The Socialists drank their beer quickly; then all would arise and go out of the tavern. The police, who had just settled down to the refreshment which they had ordered, were compelled to leave their half-empty glasses and follow the party. The Socialists strolled a short distance, then circled back—sometimes they would merely walk around the tavern and in a few minutes return to it again. All went into the beer-garden once more and ordered another serving of beer. This we called having fun with the police.

"After an hour had passed, during which we talked

on every conceivable subject except Socialism, the Socialists suddenly broke up. In a minute or two they were dispersing in all directions. The police, surprised at this turn of affairs, would not know what to do next. Consequently they would do nothing.

"But no sooner were the Socialists back in the city than they would hasten, in small groups, to a quiet hall, or a restaurant, selected beforehand. And there they would hold their meeting and listen to the speech of the Reichstag deputy, without being disturbed by the police, who had been completely outwitted."

The Socialist group in the university was now closely identified with the Social Democratic party, a movement with more than a million members throughout Germany, yet without a tangible organization, without records, without public meetings.

Steinmetz and the students quickly discovered that they were part of a remarkable manifestation of political activity. Flourishing under cover, intimidated by the authorities, outside the pale of legitimate politics, the Social Democrats nevertheless nominated candidates, conducted campaigns, and regularly elected some of their candidates to the Reichstag. And as has already been observed, once in the Reichstag, a Social Democrat was invulnerable. He could come and go openly, express his views, speak on public questions and governmental measures. Even the tyrannical Bismarck could not bring about his suppression, once his electors had mustered sufficient strength to carry the election.

Various expedients were resorted to, Steinmetz found, in operating this secret political party. Business was transacted and announcements made in a quiet manner at small group gatherings in restaurants or at cafés. Ostensibly a number of men were eating a meal together, but during the procedure some important news would be casually passed from table to table by word of mouth until all had been informed.

Sometimes at a public entertainment, at a concert, or at the theater the Social Democrats all contrived to sit together. That gave them the opportunity to engage in unostentatious discussions during intervals in the program. Out of some such furtive gatherings would come a whole slate of candidates, for which later the Socialists of the nation would cast their votes.

But the general audience at one of these entertainment occasions was wholly unconscious that a full-fledged political convention had taken place simultaneously with the program which all had assembled to enjoy.

Affiliations of this nature whetted the enthusiasm of the University of Breslau men who were pledged to the movement. The leadership of Lux was tireless; and rapidly developing as his lieutenant was the quietly loyal young student of mathematics and

electricity.   Through a full round of twelve months
they rubbed shoulders together in the "great work."

Then came real developments.   Quiet investigations which the police had been assiduously pursuing
since the Lassalle picture episode finally culminated
in 1885 with a round-up of suspected ringleaders.
The campaign went beyond the circle of university
men; some of the rank-and-file Social Democrats in
the city were also apprehended.   All told, thirty-
seven persons were placed under arrest and lodged in
jail until cases could be completed against them.

At this time Steinmetz and Hinz Lux were in very
frequent consultation.   Steinmetz was in the habit
of going to Lux's lodgings, where the latter lived
with his mother, and talking over the affairs of the
student Socialist group—as well as discussing the
movement in general.

On the day of the police campaign, Steinmetz and
Lux were chatting together in the latter's home.
Lux's mother was listening to the conversation.
Suddenly footsteps were heard in the corridor outside, followed by a loud knock on the door and a
demand to open in the name of the law.   Police
officers stalked in, rudely interrupting the conference, and announced the arrest of the zealous Lux,
who was forthwith marched off to a prison cell, leaving the surprised Steinmetz to console, as well as he
could, the dismay of the mother.

Lux and the other Socialists were imprisoned for

about ten months. Throughout that period all communication between them and the Socialists not yet apprehended was forbidden.

Nevertheless constant messages were exchanged by the two groups, messages so comprehensive that those on the inside of the prison walls always knew what was going on outside; and the reverse was also true.

The authorities had not the slightest suspicion of this. They had taken the utmost measures to prevent anything of that sort. The prosecuting attorney, who was charged with preparing the cases against the prisoners, was especially zealous in watching against violations of the edict. He was eager to make his reputation by the prosecution of the Socialists as conspirators; and his vigilance at times was grotesquely hawk-like. When he accompanied Mrs. Lux to the jail, for the brief visits with her son which the authorities allowed, he always sat between them during the conversation, to prevent the interchange of furtive notes.

Yet, as has been said, there were regular, voluminous communications, easily despatched and easily received. Lux wrote the letters sent from within the prison to those without; Steinmetz, those sent from without to the imprisoned men.

The surreptitious correspondence was made possible by an invisible writing-fluid, and the stationery consisted of the fly-leaves of text-books. It was a case where the students' knowledge of chemistry was combined with an ingenious opportunism, taking advantage of favorable circumstances to foil the government agents in a manner that must have seemed extremely uncanny to the "alert" prosecuting attorney.

At the time of his arrest, Lux was just preparing his doctor's thesis. This work he was allowed to continue during his imprisonment, the books he needed being taken to him by his mother, who also brought away the books with which he had finished. He wrote down the titles of the books he needed, the prosecuting attorney looking over the list and approving it before it was given to his mother.

When Steinmetz saw the list which was turned over to him by Lux's mother so that he could procure the books, he discovered what the prosecuting attorney could not have perceived, that Lux had cleverly worked in the symbols of certain chemicals, indicating a formula. Steinmetz immediately made some experiments and soon had hit upon a solution which could be used as an invisible writing-fluid.

The ingredients were all such as Lux was allowed to have in his prison cell. They comprised a brand of sanitary water, which he used for a mouth-wash; permanganate of potash solution, which served him as tooth-paste; blotting-paper; and a solution of hypo. The secret formula prescribed the proportion of square centimeters of blotting-paper to liquid centimeters of the fluids. The concoction that resulted was colorless and quite invisible on paper until "developed" by being held exposed to heat.

Lux and Steinmetz had an arrangement whereby if either wrote an invisible letter to the other, he marked a tiny cross in the upper corner of the fly-leaf which bore the message. When that little cross was noticed, the fly-leaf was torn out, the writing developed, the message read, and the fly-leaf then burned.

The university professors, had they been aware of the number of books asked for by Lux, would have concluded that the young man was performing an extraordinary amount of work upon his thesis. But the quantity of reading-matter that went back and forth between the prisoner and his friends did not excite the suspicions of the prosecuting attorney, although in truth it represented several times what most university students required to compile an essay which would qualify them for a degree.

Lux, with the essential assistance of Steinmetz, also used the system for a secondary purpose. He sent letters by this means to his fiancée, a Breslau young woman, and received her replies in the same manner by "return mail."

This procedure naturally placed Steinmetz in the romantic rôle of a medium of the affections, since it was he who first "developed" Lux's letter, copied it, and delivered it to the girl; then took her reply, copied it in the invisible ink upon a fly-leaf, and sent it along to Lux by means of the latter's mother. In this undertaking he served the cause of Cupid and that of friendship equally well, to the mutual satisfaction of lover and sweetheart. He virtually made it possible for love and politics, in effect, to "find a way."

Slowly the painstaking prosecuting attorney proceeded to work up his evidence for presentation before the court. One of his favorite methods was first to question the prisoners and then to ask the same questions of their friends outside. So positive was he that collusion was impossible that if, on comparing their answers to the same set of questions, both groups were found to have said the same thing, it was considered convincing evidence that both had told the truth.

Fortunately he allowed a sufficient lapse of time between his examination of the two groups to enable the invisible correspondence system to function admirably.

No sooner had he finished questioning the men in jail than Lux wrote a full report of the proceedings with his secret ink. Soon it was delivered to Steinmetz by Mrs. Lux—who was a perfectly unwitting messenger throughout these operations—its contents read, and the information made known to all members of the organization in Breslau. Then, upon being summoned before the prosecuting attorney and being asked the same questions, they were able to give answers which checked up admirably with those previously given by the men in jail—of whose testimony they were supposed to be totally ignorant! The system worked so smoothly that the authorities could make scant headway in building up a plausible case.

When at length the document of proceedings at court against the thirty-seven defendants was completed, it was shown to the men who were incarcerated, but every precaution was taken to prevent it from becoming known outside the prison. These measures were utterly futile. The proceedings formed the contents of one of Lux's letters to Steinmetz, and the following week the entire document was published in the official organ of the German Social Democratic party, to the amazement of the authorities.

The most determined of official prosecution could hardly do credit to itself against methods as subtle as these; and eventually there was nothing left to do but release the thirty-seven defendants. Evidence was lacking, even in a court of pro-Bismarckian leanings. The certainty that these young men were really conspirators against the Government remained; in truth it had never departed. But to prove it—that was altogether too difficult for the sadly puzzled prosecuting attorney. He had to own himself baffled and abandon the hope of winning imperishable renown by the proceedings against the ingenious Lux and his companions.

Steinmetz, meanwhile, was gradually coming more

and more under the clouds of suspicion. Events arising from the imprisonment of his friends endangered his own welfare, for he endeavored to fill the gap in the active work of the Breslau organization caused by the arrest of Lux and the editor of "The People's Voice."

When the imprisoned Socialists were released, in the latter part of 1887, Steinmetz was so keen in the cause that he was being watched by the police, probably more closely than he realized. An interval went by during which information against him was lodged with the rector of the university. Finally, as will be seen, the time arrived when he was unquestionably in danger of proceedings; and when, as the only alternative, he was compelled to forsake his studies on the very eve of their culmination, in order hastily to depart from a politically inhospitable Germany. He and Bismarck, he decided, could no longer breathe the same air.

CHAPTER VII

WHEN the strong arm of an intolerant government invades the editorial sanctum, however much men may desire to see the conscience of the press left unmolested, the practical result is usually the abrupt termination of the publication in question. When, in addition, there is a financial burden—debts unpaid, credit destroyed—the trials of a "substitute editor" can be imagined.

"The People's Voice," published by the Socialists of Breslau in the time of Steinmetz, was in such perplexities as these in 1886 and 1887. And Steinmetz himself appeared in the self-assumed function of stop-gap wielder of the editorial pen.

For the better part of two months—the exact period was seven weeks—Steinmetz was the mainstay of "The People's Voice." He was the editorial driving force. He supplied both the brains and the energy that made a paper possible, in the face of multiple difficulties. And the final flickering out of the publication (at least under that name) was due not so much to the activity of the Government's representatives as it was to financial straits that proved

98

too much for the young student enthusiast, himself but scantily supplied with funds for even his personal needs.

This unexpected function was thrust upon him directly by the police round-up which took his friend Lux away from him. The thirty-seven Socialists who were apprehended included both the editor of "The People's Voice" and the editor of the popular scientific magazine published in Breslau. Both were friends of Steinmetz, almost as intimate as the younger and more lively Lux. Consequently it was partly in the interest of friendship that Steinmetz was led to assume charge of the plain-spoken publication, and not alone the desire to advance the fortunes of the Socialist movement.

The first thing he did as editor was to adopt a characteristic editorial policy—characteristic alike in its naïve hostility and its youth-born bluntness. It was couched in twelve words: "We don't know what the Government wants, but we disapprove of it."

It was impossible for Steinmetz to carry on his editorial venture openly, even if he had cared to risk an undisguised pose before the vigilant police at such a perilous period as followed the arrest of his associates. The laws regulating universities expressly forbade students to engage in any regular business or profession while pursuing their studies. This in itself prevented Steinmetz from placing his name on the editorial page of "The People's Voice" as the responsible editor. He was obliged to make use of a "sitting editor."

A "sitting editor" served as a mask behind which the real editorial brains of a publication might find concealment when convenient. To the readers—and the authorities—the "sitting editor" was the real editor, unless the latter was eventually ferreted out.

The "sitting editor" in the case of "The People's Voice" was a Polish Socialist, a man of meager education, whom Steinmetz induced to take the scapegoat post. He could barely write his name and could not read at all. Yet his name was officially attached to the editorials which were regularly provided by Steinmetz. Most of the time he never knew what those editorials actually said. But he understood their general purpose to be the advancement of the party, of which he was a member; and that was enough.

As for his responsibility for the utterances of the paper, it is doubtful if he ever clearly understood this. Fortunately "The People's Voice" passed from its turbulent period of life into a calmer and more docile career after a time. And the only incident which caused the police to cast a wrathful eye in the direction of the editorial sanctum concerned Steinmetz so much more than it did the unsuspecting "sitting editor" that the latter was never really disturbed in his tranquillity.

The contents of this party organ, under the editorship of Steinmetz, were collected from various quarters. Only occasionally did he write an original article. But he reprinted articles and comments freely from other publications, and sometimes got one of the students, or a Socialist friend in the city, to write an article. In addition, he gave the general news bulletins that were of interest to the rank and file of the party.

During most of the seven weeks that Steinmetz was thus engaged, "The People's Voice" served the Breslau Socialists loyally without arousing the unfavorable notice of the Government; this, too, in the face of its somewhat belligerent anti-government battle-cry. But finally there appeared an issue which precipitated furious prosecution from the police.

There were three articles in that number which were deemed politically offensive. One of these Steinmetz wrote himself. It was a vigorous defense of the rights of the Socialists and apparently was the most stinging of the three. Another of the three had been copied from some other paper. The origin and nature of the third is not known.

That issue was the end of "The People's Voice" as a Socialist publication. The paper was banned, and the issue in question was confiscated.

Then the police, probing deeper, began to question all whom they understood to be connected with the publication. The authorship of the daring article that caused the greatest uproar was the object of their investigation. They suspected Steinmetz of having a hand in the whole affair, but proof was noticeably absent. However, they interviewed the rather guileless "sitting editor"—who promptly informed the police that he wrote that article himself!

Just why he said this is not perfectly clear; but he was so very obliging in volunteering information that the police only became the more suspicious. They went further and questioned Steinmetz, who was, in reality, the writer of the article. However, Steinmetz informed them with calm assurance that he knew nothing about it; and this was all they could learn from either of the individuals who were most closely identified with the publication.

The "sitting editor" was not molested; the police realized that whoever had written the article, it was not he. They were rather less satisfied about Steinmetz. But there was no proof—so nothing definite was done.

This was all written into the intangible case against the young mathematical student. It constituted additional evidence, even though flimsy in nature, to be laid, not long afterward, before the university authorities. And meanwhile it seemed to be excellent justification for watching this zealous, serious-faced young thinker more diligently than ever.

"The People's Voice," as stated, was never seen again. It was succeeded, shortly after the imprisoned Socialists were released, by a mild, inoffensive paper called "The Silesian News."

During this period, Steinmetz also engaged in

other literary activities. The thirty-seven imprisoned Socialists numbered, in addition to the editorial genius of "The People's Voice," another publisher friend of the young university student, who, as already stated, had been publishing a popular scientific library.

This publication, to which Steinmetz himself had written a contribution, appeared in loose sheet form from week to week, the sheets later being bound, ten or twenty together, into volumes. Each volume was devoted to one subject. There was little work for the editor, as most, if not all, of the articles were contributed. The manuscripts required nothing more than editing and transmission to the printer.

Under this same management, however, there also appeared the "Breslau Fortnightly Literary Magazine," which dealt with general topics of scholarly or classical interest, and comments on events of the times. It resembled a popular literary magazine of the present day in the United States. This was an older venture than the scientific library; but, as it proved, both of these publications were exceedingly feeble financially—a weakness which Steinmetz could do nothing to alleviate.

He exerted himself faithfully to keep the publications going, although for a few weeks he had these two and "The People's Voice," all under his care at once. But he soon discovered that the publisher of the popular scientific library and of the "Breslau Fortnightly Literary Magazine" was hopelessly in debt.   He even owed money for the paper on which the magazine was printed.

Here was a situation presupposing diplomacy and persuasion in the business office, and the usual perspicacity and judgment in the editorial department. And these qualifications were all required, at short notice, from a serious youth of mathematical and social democratic proclivities!

Steinmetz addressed himself to the double-barreled task with characteristic zest.   He interviewed the printer, who, he found, was quite willing to print another issue, if he was paid for those which had already been printed.   The paper merchant, likewise, agreed to give Steinmetz paper on credit for the next number but was rather insistent that Steinmetz first settle for the paper which had been used in the previous numbers.

Moreover, various other creditors began calling at the publication office, week after week, to ask for their money.   Most of them were subjected to an ironic bit of pleasantry.   All that Steinmetz could offer them—which, however, he did offer them—was a voluminous file of back numbers of the publications. The unappreciative creditors refused to be satisfied with such evidence of a generous spirit!

Finally, after several weeks of struggle and stress, the climax appeared with the sudden arrival one morning of the sheriff, who made the disquieting announcement that all the office furnishings and equipment had been purchased on the instalment plan and several payments were then due!

"Well, my dear sir," said Steinmetz, with the utmost affability, "I can simply offer you what I have offered all the creditors of this office—a complete file of back numbers of the publication."

And forthwith he exhibited to the dumfounded sheriff the file in question. It had no different effect upon the officer of the law than it had had upon the creditors. The sheriff merely stared in amazement at this inexplicable young fellow who could not take the situation seriously, even though the establishment was in danger of an attachment.

Then he went away in disgust, and Steinmetz had a good laugh over the whole business.

With the confiscation of "The People's Voice," which was entirely suppressed during the trial of the accused Socialists, and the financial collapse of the "Fortnightly Literary Magazine" and the popular science library, Steinmetz was perforce left with little to do for his Socialist friends except to maintain the secret correspondence with Lux by means of the invisible ink. He learned about this time that the action taken against "The People's Voice" was in harmony with the general governmental attitude toward papers of avowed Socialist leanings. All over Germany such papers were confiscated the moment any pretext was found for taking action against them; and sooner or later they all broke over the bounds of prescribed political decorum in one way or another.

Bismarck's whole policy at this time was such as to exasperate the Socialist group, thereby solidifying the movement more than ever, rather than weakening it so that it could be stamped out. The confiscation of the Socialist papers precipitated a long-drawn battle of wits, in which the honors lay largely with the Socialists, since they managed to maintain their inner system of communications by means of their party organ. This official Socialist paper was published in Zurich and smuggled into Germany by an elaborate system of "under cover" distribution.

Secret agents were scattered far and wide throughout Germany, but few of them knew definitely the contents of the package of papers which they received and passed on, week after week. If caught in the act, even though the package was found to contain copies of the outlaw paper, they could truthfully say they did not know what was in the bundle.

The man who acted as main distributing agent in Breslau was never told what the bundle contained. He seldom even read the paper himself. He knew only enough to receive the package from some one unknown to him and pass it on to some one else, who was also a stranger to him by name if not by sight. He could deny to the police virtually everything that would have incriminated him.

To be caught with a copy of the paper in one's possession meant six months' imprisonment; and almost all Socialists knew this. None the less, the paper had a circulation in Germany at large of twelve thousand copies. Steinmetz himself read it regularly, then passed it on, through the furtive delivery system, to some other member of the organization.

And now Steinmetz had a reputation elsewhere in Breslau than at the university. He had become known to the police as one of the most active instigators of the Socialist cause among the students. A case was being built up against him. As yet it was somewhat shadowy, lacking sufficient proof for proceedings which could have any reasonable hope of success.

There was one immediate result, however. The police made a full report to the rector of the university, covering everything they knew concerning Steinmetz's connection with the student Socialists. They concluded with a request for disciplinary action.

The rector of the University of Breslau was a shrewd observer. He knew more about one side of Steinmetz's career than the police did. He knew that in this young chap the university had one of its most gifted students. Not a professor whose lectures Steinmetz had attended but was impressed by his mental attributes. Steinmetz had a scholastic record which had seldom been equaled at Breslau, especially in mathematics.

In short, the plain truth of it was that the university authorities cared not the slightest what political views Steinmetz developed as long as he kept up in his studies; and this latter he was undeniably doing.

Hence it is hardly to be wondered at that the police report regarding the Socialistic activities of his most capable student made no impression upon the rector, except perhaps mildly to irritate him. He did give some attention to the request that university discipline should be exerted, but he did this only to preserve a form of official courtesy toward the representatives of the law.

He called Steinmetz before him and gently questioned the young man as to his public conduct. Steinmetz replied as gently by remarking, in effect, that he did not regard his activities as particularly reprehensible. The sedate, unperturbed rector listened quietly. Figuratively he nodded his head; he, too, did not think that the young man had been guilty of anything "particularly reprehensible." And thereupon he bade Steinmetz go his way. It was virtually a dismissal of the case.

The untiring police were more relentless. They kept at work on the Steinmetz case. They kept watch of his movements. They made further inquiries, gathered fresh data, and finally made another representation to the harassed rector.

Reluctantly the latter held a second hearing, as placid and as mild as the first—and with a similar decision, from a similar motive.

Steinmetz perceived that the university authorities were not going to stand in his way. He therefore applied himself earnestly to the work of his final year, and began the preparation of the thesis which he hoped would win him his doctor's degree. Early in 1888 he completed this work and submitted it to the professors.

The thesis was accepted by the faculty, which meant that only the formality of conferring the degree remained to be performed. The work represented by the Steinmetz thesis was extensive; it entitled him to the degree of doctor of philosophy. Yet it was but the beginning of an extended period of investigation under the same general subject, the conclusions resulting from which were not published until after Steinmetz had settled in America. The title of the thesis, translated, was, "On Involutary Self-reciprocal Correspondences in Space Which Are Defined by a Three-Dimensional Linear System of Surfaces of the nth Order."

Steinmetz never received his university degree at Breslau. On the very threshold of successfully concluding a notable university career, he was forced abruptly to relinquish the honors which he had richly earned and precipitately flee the country.

His old enemies, the police, were the cause. They had never left his trail; and now, in these early months of 1888, they had collected so much tangible evidence against him that they believed the time had

come to take action. They called in the public prosecutor, and a formal case was prepared.

Just as they were on the point of placing the young
man under arrest, a few of Steinmetz's friends managed to learn what was afoot and gave him a hasty
warning. They advised him to stand not upon the
order of his going, for to linger would mean undoubtedly a legal conviction and a lengthy term of
imprisonment.

At the same moment, other friends verified these
representations. The public prosecutor had notified
the rector of the university to abandon whatever proceedings had been instituted against Steinmetz, as
the Government was ready to proceed on its own
behalf.

Thereupon there were aroused some deeply rooted
feelings. They symbolized the freemasonry of scholarship, the bond that made brothers of professors
and students in extreme instances, like this case of
Steinmetz. The professors of Breslau, who had felt
their admiration stirred by the work of this exceptional young man, were unwilling to see so excellent
a student prosecuted and imprisoned. They contrived, in some indirect manner, to inform Steinmetz
of the notification which had been received from the
public prosecutor.

Steinmetz perceived that in truth he was a marked
man. He consulted briefly with Lux and immediately made preparations for a swift departure.

That night was his last in Breslau; his last, indeed,

in Germany. It was spent by the student Socialists in a farewell to Steinmetz, whom they all admired for his readiness to give up much rather than surrender the honest convictions of his heart.

"We all went to Lichtenheim's," said Dr. Steinmetz, recalling the occasion. "There we had one of our typical student parties, only it was more quiet than some of these affairs, and we made our revelry a little shorter than usual.

"Very early in the morning, before it was dawn, I came home, aroused my father from sleep, and told him I was going off to visit a friend and had to catch an early train. We said good-bye briefly, and without great formality. He supposed, naturally enough, that I would return after some few days, or perhaps a week. I knew I would probably never return, although I did not intimate this to him.

"After I arrived in Switzerland, I wrote to my father, telling him the real reason why I had left Germany, and informing him that I did not think I could return for a long time."

Before daylight on that May morning of 1888, young Carl Steinmetz, traveling on one of the very railroad lines of which his father was an employee, had put Breslau many miles behind him and was rapidly approaching the Austrian frontier.

CHAPTER VIII

PHILOSOPHICAL calmness manifested itself early in the life of Dr. Steinmetz. The intensive study of advanced subjects, the calculation of mathematical problems, the habit of meditation upon the ever-present social issue, all were conducive to this attitude of mind, toward which he was naturally inclined by disposition. And now the effect was to be seen in the way he went through this first great upheaval in his affairs, this hasty, covert departure from home and country.

Although, at the time, it came upon him suddenly, young Steinmetz did not become overexcited at the prospect of an unexpected—but not unforeseen—flight to the frontier. He preserved an unaffected demeanor in bidding farewell to his father. It was so completely unaffected that his father accepted his explanation of that early morning departure as perfectly natural.

But underneath all the stoic indifference was keen regret. Steinmetz would much rather have stayed in Breslau. He felt the momentous aspect of his action, realized acutely what he was doing, as he went to the railroad station and purchased a ticket for a

112

little town near the Austrian frontier where there lived a young friend of his who was a tutor.

As a matter of precaution he purchased a return-ticket, knowing as he did so that he would never use it. The ruse was successful, however; at least, his old enemies, the police, did not discover his flight until it was too late.

Dr. Steinmetz recalled this episode with an engaging frankness which revealed every particular of the naïve little plot by which he was enabled to escape from the authorities.

"The friend whom I intended to visit was a young clergyman, who had recently completed his ministerial studies. While waiting to secure a church, he was tutoring for some wealthy people in this little frontier town. He had invited me many times to pay him a visit, but I had not found it possible before to take the time to do so. Now, however, I found it suddenly extremely convenient to accept the invitation and to start without a moment's delay.

"When I got to his home, my friend quickly grasped the situation. He at once arranged an ostensible excursion to a popular summer resort a short distance across the Austrian boundary. We both bought return-tickets, but I did not return with my friend; he went back alone. I contrived to continue my journey, went to Prague, and from there to Vienna.

"I stayed in Vienna only a day or two and then journeyed on to Zurich, in Switzerland. There I decided to stay for an indefinite period. I knew there was a strong Socialist group there, and I had hopes of studying in the Zurich Polytechnic School and of earning scme money while in Zurich.

"After I had been there for a short time, I wrote to my father, telling him the real reason for my leaving Breslau, and informing him that I did not intend to return, in fact, could not return without the risk of being placed under arrest. My father did not chide me because of my course of action. He was not personally interested in politics at any time; he was not a Socialist. But he allowed me perfect freedom of thought and action. We exchanged several letters while I was in Zurich, but most of them were brief messages—news of the family on his part, and information about my studies and plans on my part. I wrote him also several times with regard to matters relating to my efforts to enter the Polytechnic."

All this time the elder Steinmetz was the only person in Breslau who knew the actual whereabouts of Carl Steinmetz, although eventually Lux received letters from his friend, which he carefully guarded.

The chagrined police were very much in the dark. Later, as will be seen, they learned of Steinmetz's presence in Zurich; learned of it in such wise that they had an excellent opportunity to thwart for a while one of the young student's ambitions. But at the moment they had to confess themselves outwitted. They were compelled to drop the proceedings which had been elaborately prepared against not only young Steinmetz but two others of the group as well, the latter having also disappeared at a peculiarly opportune moment—for them. Soon afterward, the descriptions and photographs of the men who were wanted appeared in the German governmental newspapers; but the men were never brought to trial.

From his one year in Zurich, Steinmetz gained the benefit of six months or more of study in the Zurich Polytechnic School. He there took up mechanical engineering, the most valuable addition to his technical training which this period produced. It was gained in the face of discouraging difficulties in relation to his acceptance as a student at the Polytechnic. Red tape was the great obstacle here. To be sure, it was woven about his pathway as a direct outcome of the informal manner in which he had betaken himself out of his home city; but it was none the less vexatious and had all the appearance, on the surface at least, of a mere technicality.

The trouble centered around the demand by the Polytechnic authorities for a *Heimatsschein,* or certificate of residence, which a student coming from another city was expected to secure from the police of his native town. Under the municipal law of Zurich, this was indispensable to gain entrance to the Polytechnic. Steinmetz was informed by the rector of the Polytechnic that the law was strictly enforced and that no exceptions were ever made.

The young student went away somewhat despond-

ent. He realized, of course, how high he stood in
the estimation of the Breslau police! It was not to
be expected that they would grant such a certificate
to the man whom they were about to arrest when he
left town. Yet the rector of the Polytechnic sternly
declined to accept Steinmetz as a student without it.
Moreover, as long as he stayed in Zurich as an un-
registered student, and without a *Heimatsschein* to
show, he was liable to a municipal fine, which was collected weekly by the police.

Here was a dilemma, indeed! But Steinmetz
boldly resolved to attack both horns of it at once.
He determined to make an effort to secure the *Heimatsschein* from the Breslau police, and at the same
time to spare no exertion toward winning an entrance
into the university, with or without the bothersome
document.

He at once despatched a letter to his father, asking the latter to make application to the police for
the *Heimatsschein*. Then he sought the assistance
of several friends in Zurich whose influence might
override the cast-iron ordinance, notwithstanding the
lack of the certificate.

He gained an introduction to one or two Zurich
officials, including the commissioner of police, but
none of them at first were willing to recommend that
the law be set aside in his favor. Finally a friend
introduced him to a prominent political leader, and
the latter gave him a recommendation addressed to
the rector of the Polytechnic, suggesting that it was

quite feasible to take Steinmetz in as a student notwithstanding the lack of the *Heimatsschein*. With this he called upon the rector again but was again turned away.

To add to his discomfiture, a letter from his father informed him that the Breslau police refused absolutely to grant him a *Heimatsschein*. Still he would not give up the fight. He replied to his father's letter suggesting that some good friend in Breslau might intercede for him. Perhaps he hoped that the university officials would come to his help, or that the police, perceiving they were rid of him for good, would, under those circumstances, be just as glad to make it easy for him to stay away.

All this time a Zurich police officer kept calling upon Steinmetz, week after week, to inspect his *Heimatsschein*. But every time the harassed young student had to admit that his *Heimatsschein* had not yet arrived. Thereupon the officer collected the fine prescribed by the ordinance; and the fine, Steinmetz had reason to believe, went into the officer's pocket.

The weeks passed and still no certificate. Steinmetz prayed within himself that it might appear; while the policeman, he suspected, prayed that it would n't!

In the face of all this the young man did not relax his efforts to gain admission to the university without the *Heimatsschein*. Even though there seemed to be no possible way, he did not give up. And at length he made the acquaintance of a newspaper publisher who was a considerable power in the city. This man took an interest in the student's plight; the outcome being that pressure was brought to bear, behind the scenes, upon the police, and through them upon the rector of the university.

Steinmetz never quite knew how it was accomplished. At all events, he was told to call again upon the rector. He did so, hardly expecting that his reception would differ from those he had previously received. To his astonished delight, however, the rector's attitude toward him was completely changed. He welcomed him with a smile and a bow, informed him that he might register as a student, informed him as to the necessary routine—and never once made mention of the abhorred *Heimatsschein!* Steinmetz wisely asked no questions but forthwith took up his studies with all the eagerness of a mind that was literally athirst for knowledge.

The weekly calls of the police officer and the weekly toll of fines ceased forthwith. It was fortunate; for the financial drain upon the young man's very slender resources was becoming alarming, and the levy might have continued throughout his residence in Zurich, as the certificate was never granted by the unrelenting police of Breslau.

During most of the year that Steinmetz spent in Zurich he lived in most humble circumstances. Some might have looked upon him as an impoverished youth, a typical struggling student. Such he was, so far as material means were concerned. But his circumstances did not prey upon his mind in the slightest; for Steinmetz, as a youth, was well supplied with exuberance. A serious thinker, he yet was jovial by nature, easily laying aside the consideration of scientific subjects or of political events for a thoroughly care-free evening of good fellowship around the board.

It is quite true that he lived rather precariously most of the time he was in Zurich. When he arrived there his funds were greatly depleted. He engaged modest lodgings and immediately paid for his room in advance, to make sure it was paid for at all. This seemed to be an unlooked-for procedure, judging from the surprise of the landlady, who usually had to importune her boarders with much persistence to settle their bills.

His room rent paid, Steinmetz found he had hardly any money left, but nevertheless he resolved upon the rôle of generous host toward one or two acquaintances whom he had already made among the Socialists of Zurich. He invited these men to his lodgings, providing for their entertainment a fine repast of steak and sausages. It took virtually all his remaining funds; and on the surplus food from this banquet Steinmetz managed to live for several days afterward!

Very early during his residence in Zurich he called at the home of a Dr. Simon, to whom he had been recommended. He was cordially received there, his Socialist leanings winning him a ready welcome. Among the company that he found assembled at the Simon residence was a bright young fellow who was numbered among the Socialists of the city, and who was also a student at the Polytechnic.

This student's name was Oscar Asmussen, a native of Denmark, who had been sent, when quite young, to live with a wealthy uncle in San Francisco. The uncle, seeking to give the youth the best possible start in life, had despatched him to Zurich to gain a well-rounded education. He was in the process of doing so; also of falling deeply in love with a vivacious young woman of the Swiss metropolis.

Asmussen and Steinmetz became fast friends before they had parted from each other that first evening at Dr. Simon's. Asmussen helped Steinmetz find a room and, yielding to the warm-hearted suggestion of the young Breslau student, agreed to share it with him as room-mate.

In this arrangement, Steinmetz's insatiable love of human society, his finely developed social instinct, came prominently to the surface, just as it was destined to do on recurring occasions during his later life. It was inevitable that he should seek companionship in his lodgings or his home surroundings; and this bachelor partnership with Oscar Asmussen was a strong determining factor in his life at this period, as will presently appear.

The new lodgings in which Steinmetz established

himself, with this congenial chum, were located, according to his own description, "on the top floor of the last house at the end of the last street at the edge of the town."

These were busy days for the young stranger in Zurich, who generally surprised and attracted those who met him by reason of his alert, quick manner, his never-failing pleasantness of demeanor, his strong, cordial hand-clasp that seemed to contradict his dwarfed and crippled body. He seemed to have a perfectly inexhaustible perseverance. As has already been shown, he persisted in his endeavor to attend the lectures on mechanical engineering at the Polytechnic until he gained his purpose; and he was no less indefatigable in securing the means of earning a living.

He called almost immediately upon the editor of one of the Zurich newspapers, presenting a letter of introduction and receiving a friendly hearing. The editor promptly arranged with him to write a series of articles on astronomy for the Sunday edition of his paper, the remuneration for which amounted to the equivalent of two dollars for each article.

About this time, also, the editor of the popular science library which was being published in Breslau at the time of the police round-up of Socialist leaders paid Steinmetz for his contribution to that series. This was one of the publications which Steinmetz attempted to continue for his editor friend while the latter was under arrest.

CHARLES PROTEUS STEINMETZ

The contributions of Steinmetz had been on the subject of astronomy. After appearing in the popular science series, these papers were published as a volume, constituting the first book that Steinmetz ever had published. The publisher had moved to Dresden after the ill success of his Breslau venture. It was from that city that he communicated with Steinmetz at Zurich, paying him a monthly royalty of fourteen marks, or about $3.50, from the proceeds of the book on astronomy, which began to find a considerable sale in Germany. Steinmetz also earned odd sums from time to time by tutoring.

In this semi-adventurous, Bohemian fashion, living on a very slim margin from week to week, and studying mechanical engineering in all his spare time, Steinmetz lived in Zurich, with Asmussen, his roommate. Their fare, as may be supposed, was exceedingly simple. It contained three staple items—coffee, bread, and frankfurters, the young men referring to these latter as "doggies." It was a rare event when they had butter with their bread!

One of the chief luxuries of his life was acquired by Steinmetz during this interval. It was even more of a luxury then than at any later time. And it was so inseparably a part of his daily life ever since that every one who knows about Steinmetz also knows about this habit of his. It was nothing more or less than a keen enjoyment of cigars! The habit grew upon him gradually; and he never looked back upon this year in Zurich without mildly marveling that he was able to afford cigars at all. Yet it was not long before he found a cigar more and more indispensable to his ease of mind whenever he was engaged in mental work. It was a year or two before he became such a steady smoker as to be seen at all times with a stogy; but for many years, at the latter end of his life, his cigar was so much a part of him that to see him without one did not look natural. They were not, however, "big" or "thick," as the melodramatic newspaper reporter would have us think; instead they were noticeably long and thin. And he smoked them down to a very small stump!

In the spring of 1889, after Steinmetz had mingled more or less constantly with the Socialists of Zurich, meeting among them Carl and Gerhart Hauptmann, two brothers who were at that time distinctly prominent in the city, a circumstance developed in the life of Oscar Asmussen which profoundly affected the fortunes of the young man from Breslau. This was the train of events which followed upon Asmussen's action in acquainting his uncle in California with his romantic leanings.

In describing all that happened at this time, Steinmetz exhibited a benign complacency toward the headlong pace at which his friend went about his wooing. He could well afford to, for it was the cause, indirectly at least, of his coming to the United States.

"Asmussen was deeply in love with this young lady whom he had met since he came to Zurich." So

Steinmetz recalled the episode. "He finally got to the point where he wrote to his uncle and told him all about it, mentioning also that the young lady came from a wealthy and aristocratic Swiss family. His uncle wrote back a letter of indignant disapproval. He emphasized that he had sent Asmussen to Zurich to study, not to fall in love; and he ordered him to return to San Francisco at once, enforcing this command by promptly cutting off his allowance.

"So Asmussen was compelled forthwith to make plans for going to America. He and I talked about his prospective trip a great deal. It caused us both to think about the western world, and it made me think that perhaps I'd like to go to America some day also. In fact, he and I agreed that as soon as I was able to secure funds I was to follow him.

"Then suddenly, a few nights before Asmussen was to start, as we were discussing the matter, which we were always doing then, Asmussen exclaimed: 'Well, why not come along with me?'

"'With you?' said I. 'But—I have no money.'

"'Don't mind that,' he said. 'I have enough to take us both over if we travel cheaply. I will pay our way, and you can make it up to me later.'

"It seemed like an exciting adventure to travel across the sea to the New World. And so, after thinking it over a little while, I accepted his invitation. We had a celebration in honor of this decision, with a dinner and a party that lasted until very late.

"The next morning, about ten o'clock, I had not yet got out of bed when I heard a loud knock at the door of my room. I called out for my visitor to enter, and thereupon I found him to be a Mr. Uppenborn, of Munich, the editor of a German electrical publication, in which I had written articles. He was the head of an electrical testing bureau and a scientific man of prominence. Having come to Zurich on business, he had decided to look me up because of the articles I had written for him.

"I was glad to see him and asked him to be seated. But when he looked around for a place to sit down, he found every chair in the room was filled with something, either clothes, books, or papers. He found a seat at length, however, and we had a long talk, I being still in bed. When I told him that I was planning to leave in a few days for America, he became very much interested and finally asked me to act as correspondent in America for his publication, which I agreed to do. He also gave me a letter of introduction to Rudolf Eickemeyer, who had an electrical establishment in America."

Thus, in a most informal manner, close upon the heels of his buoyant, almost haphazard decision to cross the ocean, Steinmetz had placed in his hands an important link in the chain of occurrences that was to influence his whole career after he reached the United States.

The following day, Steinmetz and Asmussen left Zurich on their long journey. Asmussen paid his friend's way as well as his own; otherwise it would have been quite impossible for Steinmetz to make the trip at that time.

Forced to avoid travel through Germany, they set out by way of Cherbourg and Paris, thence going to Havre, where they embarked on a French immigrant steamer, *La Champagne,* traveling in the steerage. They were placed with Swiss and Austrian emigrants. The rest of the steerage was occupied by Italians. Steinmetz always remembered that voyage:

"It was the most pleasant trip I ever made. I was not once seasick, and we made a very jolly excursion of it, all the way over."

The voyage consumed eight days. During most of that time Steinmetz did his best to pick up a little English. Asmussen gave him what instruction he could, but the time was so brief that when they finally reached America, Steinmetz could speak hardly any English beyond a few brief phrases.

It was now the latter part of May, 1889. Steinmetz had been away from Breslau about a year. That move had been a decided break in his affairs, one of a succession of sharp changes that followed each other through much of his life. Yet the step he took in coming to America was a far bigger break, upon which impended happenings that he himself foresaw not.

And here he was at length, a-sail on the broad Atlantic, cutting loose from the last vestige of the world and the life that was familiar to him, going to a country whose language he knew not, dependent upon a friend and fellow-traveler for funds, and without the slightest idea as to what he was to do for a living. Yet he had his health, he was young, and he was enthusiastic.

His other assets were bigger than either he or any one else dreamed of at the moment. The course of time and the play of events were to bring these temporarily hidden talents very conspicuously to the surface in the new, unknown land where a brilliant career awaited him.

CHAPTER IX

A WARM, pleasant afternoon—the first of June, 1889. The keen sea air whiffing over the waves of the lower bay. The French immigrant liner *La Champagne* working her way through the passing sea traffic, and off ahead of her a sight for the soul to gaze upon and never forget— the great figure of Liberty, with her quenchless torch!

Many wondering eyes looked up at that figure from the deck of the French liner; eyes that had never seen the spectacle before. The young German, Steinmetz, looked eagerly with the rest. But he was frank in saying, in his later years, that he was hardly any more contemplative, at the moment, of the symbolism thus arisen before him, than were the other immigrants in their exciting realization that America was at hand.

Yes, here was America! Young Steinmetz, turning from the heroic statue, looked with equal wonder round about upon the busy scenes of the great port, the vessels looming up with the drift of smoke about them, and the first suggestion of the bustling city, the ever-astonishing waterfront of the western metropolis.

The spirit of it all took hold upon him that sunny afternoon, coming up the harbor. He felt the surge of adventure. The sights, the sounds, the smells of this new sea-washed shore were all about him. America and her advantages were swiftly imagined as he stood there. And they were inevitably more sharply defined than is the case with the young man who is born on the soil and grows up in the land, having his country's grandeur unfolded gradually before him.

But the more unpleasant realities of life, as encountered in the process of landing from an immigrant vessel, intruded upon his romantic fancies.

It was a Saturday afternoon when the vessel docked. The cabin passengers were put ashore at once. But those in the steerage, Steinmetz and Asmussen among them, were held on board until Monday. Saturday and Sunday were warm, pleasant days; but on Sunday night the wind changed, blowing damp and cold through an open port upon Steinmetz's head as he slept.

This sudden veering of the breeze, his first experience with the fickle North American climate, complicated his landing experience most uncomfortably, for he awoke with a very bad cold, which caused one side of his face to become swollen, and made him feel miserable generally.

Yet he hopefully confronted the immigration officers, accompanied by his loyal friend. Together they landed at old Castle Garden, now the Aquarium, forerunner of Ellis Island.

If Steinmetz had been alone that day, his dream of coming to America might have ended abruptly right there at Castle Garden. His forlorn appearance, swollen face, empty purse, and stumbling English caused the immigration authorities to shake their heads. His knowledge of English was so scant that when the officials asked him if he knew the language he could only reply, "A few." After some minutes of searching questions and puzzled answers the official decision was reached and made known to him. He could not land! He must go back to Europe!

The tremendous disappointment that leaped into his eyes when he understood this decree did not alter the official attitude. With disconcerting briskness they sent him to the detention pen.

But his traveling companion saved the situation. Asmussen explained to the officers that Steinmetz and he were together. He stoutly declared they would stick together after landing and that he would personally see that Steinmetz did not become destitute in a strange land. Asmussen spoke English fluently. Moreover, he showed a fairly substantial sum of money, which, he declared, belonged to them both. He was willing to make himself responsible for the welfare of his friend; and, upon his representations, Charles P. Steinmetz was finally admitted to

America, which was to be for him the land of friends, fame, and fortune.

Yet, as already revealed, Steinmetz himself had not a penny in his pocket when he first set foot upon American soil. His traveling companion was his financier. He owed his friend money for paying his way over from Europe; he owed him also for any expenses that came up from day to day. He himself was destitute.

Thus it was that the two young friends found themselves in New York with both funds and prospects uncertain. Yet those few weeks were weeks of happiness for them. Asmussen had relatives in Brooklyn, and there they obtained lodgings until they could hunt up work.

The Brooklyn people received Steinmetz with as much hospitality as they did their own kinsman. They made it pleasant for him while he boarded there, and helped him to learn English better. And soon he began hopefully to visit the places of which he had been told, or to which he had letters of introduction, looking for a chance to work.

The first person whom he approached, seeking a position, was the engineer of the Edison machine works, to whom he bore a letter of introduction, written by Mr. Uppenborn. But there was no opening here; the engineer made that quite plain.

"It seems to me," he remarked, as he dismissed his caller, "as if there was a regular epidemic of electricians coming to America."

CHARLES PROTEUS STEINMETZ

The next day Steinmetz went to Yonkers and called upon Rudolf Eickemeyer, who conducted a prominent manufacturing establishment near the railroad station, having succeeded the original firm of Eickemeyer & Osterheld. What happened when he entered the office is told by Walter S. Emerson, a nephew of Eickemeyer's, who was at that time an office clerk, in addition to other duties.

"He had come directly from the railroad station," Mr. Emerson relates. "He wore plain, rather rough clothes and a cap. I got the idea, from looking at him, that he was some chap who had knocked his way from place to place, looking for a job.

"I asked him whom he wanted to see. He replied: 'Mr. Eickemeyer,' speaking in a quick manner.

"I went up-stairs and found Mr. Eickemeyer. 'Uncle,' I said, 'there's a man to see you down in the office. I don't know his name; he might be a fellow who has come off a freight-train. I'll follow you down.'

"I went down behind Mr. Eickemeyer and stood in the door as the two met. Then I heard the visitor's name. I heard him say: 'I'm Mr. Steinmetz'; and then they began to talk German and sat down together at Mr. Eickemeyer's desk.

"I stayed a little while, then left. A little later I glanced into the office. They were still talking together, Mr. Eickemeyer sitting at his desk and Steinmetz in a chair alongside. They talked for a couple of hours."

STEINMETZ AN AMERICAN

It seems that when Eickemeyer made his appearance, Steinmetz arose and said, mustering his best English:

"Have I the honor to speak to Mr. R. Eickemeyer?"

Eickemeyer, who was a good judge of men, looked the young fellow over with a quick, keen eye, nodded understandingly, and said, with a smile:

"Sprechen Sie Deutsch?"

And then all went well. Feeling more and more at home, Steinmetz talked eagerly with Eickemeyer in German. They enjoyed a stimulating conversation, in the course of which all the latest news in electricity and electrical matters, as well as the most recent technical developments, were fully discussed.

Steinmetz himself pictures this notable meeting as an opportunity for both to engage in an enjoyable technical discussion about various electrical subjects.

"I inquired if I had the honor to be addressing Mr. Eickemeyer, whereupon he said, 'Sprechen Sie Deutsch,' uttering it as a command, or a request, rather than a question. I then presented my letter of introduction from Mr. Uppenborn., After that we talked for an hour or more, our chief subjects of conversation being transformers, storage-batteries, and apparatus having to do with magnetism. He was much interested in the subject of transformers and storage-batteries, inquiring from me the latest developments in Europe concerning these machines. He was doing but little real electrical work at that time, his business consisting chiefly of the manufacture of hat machines."

This interesting interview did not produce a position for Steinmetz, however. All that Eickemeyer could do was to take the young man's name and address, promising to inform him if an opening occurred.

But Carl Steinmetz was not the sort of fellow to sit down and wait for opportunity to seek him out. A week later he again presented himself at Eickemeyer's plant, to see if there was a chance for him. His persistence was rewarded. He was told to report for work the following Monday morning.

His job was to be that of a draftsman at two dollars a day, or twelve dollars a week. And that was his start in America, secured principally by his persistent effort, within two weeks after he landed at Castle Garden.

A profitable commentary on the turn of events which made Steinmetz an employe of Rudolf Eickemeyer is to be found in the "General Electric Review" for September, 1912. Writing under the title of "Steinmetz and His Discovery of the Hysteresis Law," Douglas S. Martin says:

"Two weeks later [after landing at Castle Garden], Steinmetz presented his letter of introduction to Rudolf Eickemeyer at Yonkers. The element of chance probably was at work here, as there were other nebulous plans then in his head; but it may also be in the belief that here would at least be work and opportunity for a man with some knowledge of electrical matters. Actually his knowledge was somewhat limited, at least so far as practice was concerned. He had never handled—hardly ever seen—even a direct current motor; and, although he had published in Switzerland an able paper on the design of transformers, the sight of one of them ('converters' as they were then called) had never been vouchsafed to him."

Promptly upon finding employment, Steinmetz took steps to establish himself in the western republic in another respect. Unwavering in his decision that America would be his home and his country thenceforth, he had speedily appeared before a naturalization court and had taken out his first papers. He wanted to be made a citizen of the new land to which he had come. This purpose was consummated in due time, for five years later he returned to Yonkers and received his second papers, which raised him to the status of a fully naturalized citizen of the United States.

Almost the moment that Steinmetz had steady work, he dropped into an easy, Bohemian mode of living, which yet catered in a peculiar way to his peculiar social needs. In a sense it was the forerunner of a more pronounced period of queer masculine housekeeping many years later in Schenectady, out of which finally blossomed the home life that permanently enriched his mature years.

135

CHARLES PROTEUS STEINMETZ

This first adventure in rough-and-ready domesticity was shared by his chum, Asmussen (who had found it desirable to stay in the East for a while), just as his later bachelor establishment was to include his adopted son, Joseph LeRoy Hayden.

Steinmetz was always chummy by instinct. He hated a solitary life; and his disposition made it easy for him to avoid such a state, for his intimate associates could scarcely ask a more thoughtful or courteous companion. That was what first attracted Asmussen to him. And now, as chance decreed, both of them were making a start in the same city, for Asmussen had found a place with the De La Vergne Refrigerating Machine Company, also in Yonkers.

From Brooklyn to Yonkers is a long, monotonous trip to make every morning, day after day. The two young men had to rise at five o'clock, eat a lunchroom breakfast, take the Roosevelt Ferry to Manhattan, rattle up to Forty-second Street on the elevated railroad, and then get a train for Yonkers.

This was too inconvenient to last; and after a few days they rented a room in Harlem, where they set up housekeeping. At least, they called it housekeeping, although doubtless it would have shocked any systematic housewife, especially those of strict New England proclivities.

They divided the household duties as equitably as might be. Steinmetz, who got up early and took a train at 7:15, was responsible for preparing breakfast for them both. Asmussen, who was able to get home

RUDOLF EICKEMEYER OF YONKERS

earlier in the evening than Steinmetz, acted as purchasing-agent, laying in all the supplies and getting the supper.

It was a lark for them both. And they were both alert to enjoy it. Sometimes they had the most laughable, uproarious arguments. And sometimes they provoked each other to tremendous, though good-humored, protests over various high-handed proceedings.

Thus, when Steinmetz found it desirable to take a lunch with him, he nearly drained the coffee-percolator each morning to provide for his own breakfast and to fill the coffee-bottle in his lunch-box. Consequently, Asmussen, upon arising a little later, found so little coffee left that it was necessary for him to brew a fresh supply. Repeatedly the Steinmetz longing for the beverage put him to this inconvenience. But he evened up the score at night; for, being the first to get home, he always drank up the best of the tea before Steinmetz could arrive.

The dish-washing question was the issue on which they most frequently split. As in many another household, it was an eternal bane. Each tried to shirk it. And each kept strict watch to prevent the other from shirking. The upshot was a succession of comical altercations, ending as a rule in a truce, under which, by mutual consent, the dishes went unwashed until they were next needed.

After a while they conceived the excellent idea of dispensing with plates by using squares of paper, which could be thrown away after each meal. But there were still the cups—and, worst of all, the frying-pan!

They discovered, a little later, a larger and more pleasant room, on 122nd Street, and moved in. This room had a big stove, but they still continued to use gas for their cooking. The stove served a more useful purpose for them: it became a common rubbish receptacle and was almost always filled up with paper wrappings that had contained meat, cheese, butter, and other provisions.

The result was what might have been expected. One evening, when the two young men were eating supper, they suddenly discovered a mouse quietly running about, picking up crumbs. Amused at the temerity of the little animal, they threw it some bits of food. They were surprised to see the mouse, after nibbling away, suddenly run into the stove. Investigation disclosed that the stove was undoubtedly the headquarters of a good-sized tribe.

It was not long before a number of mice ventured boldly out into the room. By throwing crumbs to them on many occasions, Steinmetz and his friend finally got the mice so tame that they felt quite at home in the presence of their human benefactors.

They felt so much at home, in fact, that they proceeded to multiply exceedingly. Soon they were overrunning the place. The men stopped feeding them, and the mice began eating the food supplies without asking permission. More than that, they began dining on clothing, especially handkerchiefs.

This, of course, could not continue. Yet Steinmetz was quite unable to bear the thought of killing his tame mice. He protested against such a monstrous suggestion. Asmussen was easily persuaded that it would be quite a needless cruelty. There was a far better method of dealing with the problem. It was a perfectly simple plan, which young Steinmetz advocated, Asmussen agreed to, and both promptly carried out.

They just moved away, leaving their beloved mice to whatever fate might befall, but, for the time being at least, in complete possession of the premises.

Six weeks had now passed, and the young men had advanced in their work. Both were receiving eighteen dollars a week. By living economically they had saved a tidy sum, all of which was regarded as the possession of Asmussen, because he had paid the entire expense of their joint trip to America.

The savings which Asmussen thus acquired finally enabled him to send for his fiancée in Switzerland in calm defiance of his uncle in California. She came to the United States shortly afterward, and they were married in New York.

They rented an apartment in the Bronx, and promptly invited Steinmetz to make his home with them. Steinmetz, however, was getting more and more interested in his work. He wanted to travel to and from the Eickemeyer plant more readily. So he decided to locate near the factory, and found a place in Yonkers that suited him.

This was in the home of Edward Mueller, who was the factory draftsman at Eickemeyer's when Steinmetz first started working for Eickemeyer.

Of his life in the Mueller home there is a lively reference in a letter which Mr. Mueller wrote to him on December 30, 1911, thanking him for Christmas remembrances which Dr. Steinmetz had sent to the Mueller family. It is evident from passages in this letter that Steinmetz was very much of a fun-loving young fellow, and that he enjoyed some merry times with his Yonkers friends.

"We are deeply touched," wrote Mr. Mueller, "by the way your memory keeps fresh all the details of those good old times. Indeed do we remember those happy, interesting days, when you won your bet that you would eat all the noodles, when you gave a lecture on electricity to the German Socialist section; the time of grand Eickemeyer and old Getty, and Dick Tischendorfer; when you read your first paper on hysteresis in your rubber shoes and with your trousers turned up; the big sleigh that was built in the den, as well as the geometric soap bubbles; when you used to pull mother's apron-strings, or tie her in her chair with same, when you chased the girls around the table with a broom, etc., etc., etc."

The weeks and months that now followed constituted the first period of Americanization for Steinmetz. He began to be an American in spirit. He observed the American customs and habits that passed before his eyes in teeming succession from day to day. He saw American scenes and meditated upon them, met American men and women and studied them. Week by week his English began to improve. He gave as much spare time as he could to becoming familiar with the language, and sometimes he received assistance from the friends he was making. Before long he was able to converse with increasing ease in what had been only a few months before an entirely unfamiliar tongue.

From the moment that he became established in America, Steinmetz allied himself with the technical agencies of his profession. Within a few months he made application for membership in the American Institute of Electrical Engineers, and was admitted toward the end of 1889. He also joined the New York Mathematical Society, now the American Mathematical Society, taking an active part in this organization for a number of years and reading several original papers.

But now he was finding engineering making more demands upon him than mathematics. Slowly he began to give more attention to the former and less to the latter—not because he wanted to, for Steinmetz was always a mathematician by choice and enjoyment, but because engineering problems and developments came to constitute more entirely his daily work.

"Gradually," he says of this period, "I drifted out of pure mathematics, to my very great regret; but engineering now occupied all my time."

His first public appearance in America was at a meeting of the American Institute of Electrical Engineers in 1890. It was during a discussion that followed a paper on "The Armature Reaction of Alternators," read by Thorburn Reid. Steinmetz criticized the theory which Mr. Reid advanced in this paper as incomplete because the third harmonics had not been considered. The author of the paper challenged the criticism by stating that a consideration of the third harmonics would make the theory too complicated.

Nothing more was said upon the subject at that meeting, but Steinmetz quietly proceeded to work out the theory, including the third harmonics. Months afterward, to the surprise of Mr. Reid and the other engineers present, he unfolded this theory in the first paper which he ever read before the American Institute of Electrical Engineers.

Eventually Mr. Reid and Steinmetz became the best of friends. Mr. Reid was at one time an associate of Dr. Steinmetz in the General Electric Company and assisted him in publishing the first edition of one of his most successful technical books, "Alternating Current Phenomena."

There were other incidents during this period similar to the one just mentioned, all of which began to direct more than passing attention to the young German scientist who was on Eickemeyer's staff at Yonkers. That he was a master mathematician, although only twenty-four years of age, was evident. Men began to show a readiness to listen when he spoke, in the belief that he would have something to contribute to their engineering and mathematical knowledge.

His reputation in America was taking form. The stage was almost set for his first startling revelation of his mathematical ability.

At the same time, it is apparent that circumstances largely steered him into the field of electricity and electrical problems. He did not—indeed, could not —deliberately and consciously give up mathematics for electricity, or for anything else. On this point Martin writes with keen appreciation of the situation:

"It has become increasingly apparent that the science of applied electricity would have been a heavy loser had that slight inclination toward matters electrical been omitted from Steinmetz's composition. For he is as much mathematician as engineer, and as much physicist as mathematician; and he might so easily have decided to leave electricity alone. It is only by reflecting upon what he has since achieved and the regard in which he is held by electrical men, whether they know him or not, that we can be sufficiently thankful for the chance which landed him, twenty-four years old, on these shores on the first of June, 1889; and the chance (it was little more) which led him into Eickemeyer's factory two weeks later."

CHAPTER X

THE world of electrical men and electrical affairs was not concerning itself exclusively by any means with the river city of Yonkers in 1889 and 1890. In fact, its attention just then was much more closely concentrated upon Menlo Park, New Jersey, where Edison was doing astonishing things. Folk had not ceased marveling over the incandescent electric lamp, which was just coming into its own.

Yet in Yonkers, during that period, a young fellow who seemed but a youth compared to the great inventor—in truth Steinmetz was not quite twenty years younger than Thomas A. Edison—was quietly working and studying and, most significant of all, was thinking. And his thinking was deeply concerned with his work.

For a while this young chap was engrossed in his daily duties as draftsman. He was learning the ropes in the establishment where he worked. But that did not last long. Steinmetz, of all men, was able to master rapidly the details of a given task. Consequently, when his work began to arouse his initiative and to lead him on to become an original thinker, as it speedily did, he threw into it all the enthusiasm of a genius.

An excellent character glimpse of the man a few weeks after he started work with Eickemeyer is afforded by Walter S. Emerson, who, as a fellow-employee, sometimes saw Steinmetz about the plant.

"He picked things up mighty quickly," recalls Mr. Emerson. "I don't remember anybody I ever saw at my uncle's factory who picked up the ropes quicker than Mr. Steinmetz. He was a pleasant, genial man, whom everybody got to like. He moved about so quickly and alertly that it was a matter of general comment. He would come downstairs two steps at a time. He was up and down like lightning. If we heard a clatter of feet on the stairs we'd remark to one another: 'There's Steinmetz, coming down on his neck.'

"Not long after he came, he and my uncle became engrossed in field-coil work. I overheard them talking together one day, and caught a quick, eager remark from Steinmetz: 'We've got to excite the fields! We've got to excite the fields!' I did not understand what it meant, but I could not forget the tremendously absorbed, quick way in which he spoke.

"He was frequently doing small jobs for persons around the plant. When we were cleaning up in the offices some time later I came upon a gold watch laid away on a top shelf in his laboratory, and restored it to him. He was glad to get it and told me he had taken it to get the magnetism out of it for some one he knew and had then forgotten to return it.

"Naturally I have changed my opinion of Mr. Steinmetz from the impression I got of him the first day I saw him, when he looked to me like a young fellow who had been traveling 'blind bagagge.' I soon became aware that Steinmetz was no tramp. I have never stopped contrasting his final position and fame with that period of his first work in my uncle's establishment, when I knew him as one of us, an every-day worker like all the rest, so far as we could see. It is amazing to me to think that a man could have a brain that would carry him ahead so rapidly and so far."

The story of the magnetized watch bears upon a procedure which was doubtless duplicated upon many occasions. The immediate vicinity of a piece of electrical apparatus in those early days of modern electrical science was quite likely to be charged with magnetic currents.

All dynamos and motors of the period were crude by comparison with the Eickemeyer types, which were of an absolutely new design and confined all of the lines of magnetic force within the frame of the apparatus so that the surrounding air was free of magnetic leakage. It was therefore possible to bring a watch within the magnetic field of these machines without affecting the steel works and thus rendering it useless as a timepiece.

Magnetism could be drawn out of a watch by a

simple process, discovered by Steinmetz, restoring it to its time-keeping function; and the watch referred to had evidently been left with him for treatment, by some friend outside of the establishment, who had ventured too near an old-type machine of other makers.

It has been said that Steinmetz drifted into Eickemeyer's factory by little else than a happy chance. But it was more than mere chance; it was something very much akin, in certain aspects, to a strange leading of fate.

This does not seem in the least an exaggerated view when it is understood how completely Eickemeyer's influence brought out the best that Steinmetz had to offer; started him, in effect, upon that line of research which enabled the young mathematician to put his remarkable ability to useful service.

At this potentially momentous interval in his life, Steinmetz was all aflame with scientific enthusiasm, young and buoyant, intensely interested in life and in the great world which is the stage of all human enterprise and achievement. But he was in the rough. He had not yet found himself. He did not even know, with full assurance, what he wanted to do in life, what career would suit him the best.

He very much needed a congenial, understanding, encouraging mind to open up for him all the possibilities that were crowding close at hand in electrical science. And that was precisely what Eickemeyer provided. Speculation is idle as to what might have been; yet it is quite pertinent to raise the question as to whether Steinmetz, amid less stimulating conditions, would have been led to make the investigations he did make, or to become so swiftly and so surely a giant in electrical engineering.

Steinmetz himself, years later, spoke his solemn conviction upon this point. It was during a quiet early autumn evening at his beloved camp on the Mohawk in Schenectady, when he had lapsed into a mood of gentle reminiscence. There was with him at the time Mr. Eickemeyer's son, Rudolf Eickemeyer, Jr., who was making his first visit at the camp. The two had drifted into talk about the old days in Yonkers, when the biggest problems of electrical science had not yet been entirely unmasked. Finally Mr. Eickemeyer said:

"Why is it that you think so much of my father? What has so exalted his memory in your mind?"

"I 'll tell you, Rudolf," answered Steinmetz, "why your father's name is dear to me. I came to him with a vast amount of knowledge which I had accumulated through the years. It did not enable me to more than make a bare living. Your father took me in hand, made me a part of his intimate business relationship in the line of his inventions, and showed me as long as I was with him how I could apply my knowledge and make myself useful to myself and to the world."

ELECTRICITY AT YONKERS

Rudolf Eickemeyer, Sr., of whom Steinmetz thus spoke so feelingly, and to whom he later dedicated his first scientific text-book, was unmistakably a man of remarkable scientific ability, as well as a leading inventor. He had revolutionized the hat industry, through his improvements in hat-making machinery, and later brought out many epochal inventions of an electrical nature, including the first alternating current motor, the use of carbon for brushes, a method of sectional windings for armatures which greatly improved this process, a new iron-clad dynamo and motor, and his ingenious magnetic bridge. His total inventions numbered more than one hundred and fifty.

Eickemeyer's early career was similar, in one or two episodes, to that of Steinmetz. He, too, had been involved in agitation against the German Government. And he, too, had left Germany a refugee. He was a revolutionist in the revolution of 1848, as an outcome of which thousands of fine young Germans migrated to America. It was the first stirrings of a new spirit in Germany, which was to struggle for years against unsympathetic governmental forces in the trend toward modernism.

At the time of the revolution Eickemeyer was a student in South Germany. The students were a prominent factor in the uprising, joining the movement in large numbers. It required the presence of Prussian troops, with uncompromising orders to "shoot to kill," to quench the flame of the revolutionary activity and lead the students to return to the universities.

The universities, however, received them but coldly. None were actually expelled, but life was made so unpleasant for them that many took refuge in an enforced immigration.

Eickemeyer was one of these. He and his chum, George Osterheld, scraped together all the money they could and came to America. They got work as laborers on the New York Central Railroad, then in course of construction, at Lodi, near Buffalo. Later Eickemeyer's mechanical ability got him a place in a large shop in Buffalo, where he one day led a strike. The strike was successful; nevertheless, Eickemeyer found it advisable to get work elsewhere after the trouble was over. This was what caused him eventually to make his way to Yonkers, there to establish a small machine shop.

As Yonkers was then the center of the hat-making industry, Eickemeyer quickly found plenty to do repairing hat machinery. He became interested in the hat machines themselves, made improvements in them, and took out patents. Then he began manufacturing his improved machines.

A few years later, electrical developments began to attract Eickemeyer's attention. Seeking diversion by a different type of endeavor, he went into the electrical field in a tentative, half-inquiring fashion. He conducted a number of experiments, which assumed such serious import that he followed them with many successful inventions.

Martin has the following to say about Eickemeyer, in his writings in the "General Electric Review":

"Apart from the extent to which the electrical industry as a whole is indebted to him [Eickemeyer] for pioneer work in the design of alternating current machines, Steinmetz himself owed much to the inspiration which he drew from close contact with the older man in his researches on magnetic materials.

"Well as Eickemeyer's name is known amongst electrical engineers, his work in this field was in reality no more than incidental to the main work of his life. Hat-making may sound like an unromantic calling; but in the middle of last century, when the crude and wasteful methods of hand production represented a condition which was crying for a man who could produce automatic apparatus, the hat business as a vocation was full of attraction for a man with imagination and the ability to invent. Eickemeyer possessed both; and they enabled him to revolutionize the then prevailing practices in the hat-making industry, both in America and abroad."

A further insight into the dynamic career of the one man who did more than any one else to develop Steinmetz is found in the "Electrical Engineer" for December 17, 1890:

"He had always followed in a general way the various advances in the sciences, taking great interest in electricity; and when the Bell telephone was brought out, having more time and means at his disposal than he had previously enjoyed, he took up electricity as a study for his leisure, rather than with a view to applying it in a practical way. Experimenting with various forms of telephones, he became familiar with the peculiarities of different forms of electromagnets; and from their use in telephones to the construction of dynamos was but a step.

"The celebrated ironclad dynamos and motors known as the 'Eickemeyer' were the first attempt on his part to put to practical use the results of electrical studies and investigations extending over a space of nearly ten years. . . . The Eickemeyer motors have also been applied to electrical railway work, and found highly efficient.

"The question of the best material to be used in the construction of dynamos caused Mr. Eickemeyer to set to work to get some instrument that would enable him to determine readily the relative values, magnetically, of various qualities of iron and steel; and the result was a magnetic bridge, by means of which the magnetic value of the material could be told as readily as a loaf of bread is weighed on the scales of a bakery. The instrument has proved a complete magnetic laboratory in itself, and by its use Mr. Eickemeyer has been able to determine many questions in the construction of electric devices which, without it, would have been difficult and expensive experiments."

ELECTRICITY AT YONKERS

Steinmetz gave Eickemeyer unreserved credit for these same achievements, which he came to appreciate the longer he was associated with this valuable friend.

"The original partner of Mr. Eickemeyer was George Osterheld," Steinmetz told his adopted son, J. LeRoy Hayden, a few months before his death. "Upon his death, a younger brother, Henry Osterheld, was made a member.

"Mr. Eickemeyer took a great interest in all my investigations. In fact, our relations quickly became very cordial. It was an establishment with a little over one hundred employees, and Mr. Eickemeyer was pretty much the whole supervisory force. He soon found out I could calculate and thereupon asked me to do that sort of work. So I went into work on magnetism, and he allowed me to operate his magnetic apparatus, including his magnetic bridge.

"Studies in magnetism had been quite extensively made by Mr. Eickemeyer before I went into his organization. I carried them on further, working at this all through 1891 and 1892, under Mr. Eickemeyer's direction. While I was in Switzerland, I had studied transformers, which were new in 1888. I wrote a paper on the theory of alternating current transformers, which was published in German; but I did not actually see a transformer until I got to Yonkers. I then wrote another paper, further studying this subject."

CHARLES PROTEUS STEINMETZ

When Steinmetz went to work for him, Eickemeyer was one of the leading citizens of Yonkers.
His home was noted for its delightful open-hearted
hospitality. A large, square, handsome house, built
of brick, it was set high on a commanding bluff on
Linden Street and was known as Seven Oaks. A
grove of century-old trees, which fairly surrounded
it, was the origin of the name. There were many
servants, and large, commodious rooms, for Eickemeyer was a man of prosperity. His family consisted of his wife, three boys, and three girls.

As he had begun his electrical experiments only a
short time previously, Eickemeyer was then manufacturing a comparatively small number of motors
and generators (or dynamos, as they were then
called), and these were of a very early type. They
were all designed for direct current operation and
were to be used for all purposes, including electric
cars. This latter was an application of electricity
in which there was just then much experimental activity but as yet no practical usefulness, as the experiments had not met with success up to that time.

One of the pioneers in this undertaking, however,
Stephen D. Field, a nephew of Cyrus W. Field, had
come into Eickemeyer's establishment shortly before
the appearance of Steinmetz. Field was continually
endeavoring to develop an electric railway car; in
fact, he had already devised a crude type of electric
locomotive, which was originally built for the New
York Elevated Railroad Company, although it

failed to come up to expectations in actual test.

Field had been associated with Eickemeyer since 1884, having in that year brought his design for an electric car motor to Eickemeyer, who was engaged in building the machine. The motor was covered in a number of patents, which were held by Field; and in seeking to have it constructed for actual testing he made the Eickemeyer factory his manufacturing agent. Eickemeyer and Field eventually organized a separate company for exploiting their combined street-car motors, but they never reached the manufacturing stage, on account of the sale of the Eickemeyer plant to the General Electric Company. Some of Field's first motors were the motive power on New York's first cross-town electric cars, until supplanted by more improved types.

Monday, June 10, 1889, is the date on which Carl Steinmetz entered the employ of Rudolf Eickemeyer. This is clearly established by Steinmetz's time-book, neatly and faithfully kept in his own handwriting. On the cover it bears the words "Time Book," also in Steinmetz's painstaking hand, with its quaint German script, and above appears the label, affixed with a rubber stamp: "Charles Steinmetz, Electr. Engineer."

It can be seen from this that Steinmetz had begun to Americanize his name, but had not yet inserted the initial "P." in place of his two middle names. The entire four years which Steinmetz spent as an employee of Eickmeyer's are accurately recorded in this time-book. The particular piece of work on which the young draftsman was engaged each day is put down, and every holiday is noted. There are also two or three periods, of several days' duration each, where the notation opposite the dates reads "Sick with la grippe." The record, in one way or another, accounts for every working day of the entire period.

It also discloses that the greatest mathematical and electrical genius of his age began work with a ten-hour day. Every one has heard of Steinmetz's famous prediction, uttered during the last year or two of his life, that within the next century the industrial world would operate on a four-hour work-day, which, he believed, is as long a period of continuous labor as men ought to undergo. He himself knew what it meant, in his own life, to put in more than twice that length of time as a working-man; but he also knew how little mere hours count when men's work fascinates them to such a degree that it ceases to be work at all, in the sense of toil. Whatever his views in those first years in America, he was as reliable as the sun itself in his own work, for his time-book indicates that he had a full ten hours to his credit every day of his Yonkers employment, except when he was absent because of sickness.

His first drafting job, on that tenth day of June, 1889, was to make an assembly drawing of Eickemeyer's detailed drawings of the motor for electric cars. This work he noted in his time-book as "street car motor No. 3."

He was busy with this until June 22, on which date he began to make drawings for "switch of overhead conductor"; and this continued until July 6, when his work changed to drawings for "electric trolley." All of these drawings were related to the proposed electric street-car.

From July 11 to July 13, he worked on an electric water-pump, and from July 13 to 22 he was busy with fourteen-inch and twelve-inch dynamos; then, until July 22, with a drawing for an alternating current motor, all these being Eickemeyer's devices.

And this last is significant in itself, since it foreshadows the work which precipitated that notable line of investigation resulting in his discovery of the law of magnetic losses in alternating current motors, the thing that first made Steinmetz famous among electrical engineers.

Affixed to the last page of entries in the time-book is a receipt for fifty dollars for "wages in full to date," signed by "Charles P. Steinmetz," and dated February 15, 1893. The "P." here appears in his name, and thus it is clear that he adopted his unique middle name of "Proteus" some time between 1889 and 1893.

The way in which this came about, as Dr. Steinmetz himself used to relate it, was that, having become an American, and feeling proud of the fact, he wanted to have a truly American name. Hence he called himself at first plain Charles Steinmetz. But after a while he observed that the names of most Americans consisted of a first name and a middle initial. So, to be like other Americans, he resolved that he must have a middle initial, also. He did not immediately decide what the initial should be and was still pondering the matter when an old friend of his university days in Germany dropped in one day to renew acquaintance. He greeted Steinmetz with a boisterous "How are you, Proteus?" and a warm grasp of the hand. The sound of the old term awoke pleasant memories in Steinmetz's mind, stirring his ever-susceptible social trait. He resolved in that moment to perpetuate the old university nickname by incorporating it into his new Americanized name; and from that time forth he became Charles Proteus Steinmetz, commonly signing himself Charles P. Steinmetz. And this, it must be admitted, had as truly an American aspect as could possibly be desired.

From the outset, Steinmetz found plenty of room for the exercise of his initiative in assisting in the making of finished drawings of the electric car motors which had been sketched by Eickemeyer and Field. The same can be said of his work on the detailed drawings of Eickemeyer's original line of motors and dynamos which were forthcoming at a slightly later period.

The investigations of these two men, each a genius

in his own way, soon dropped into a permanent relationship. Eickemeyer at that time was displaying truly remarkable inventive ability; and Steinmetz assisted in perfecting the working drawings of these various inventions. In doing so, it seems quite likely that he observed possibilities for going a step further, a change in this detail or that, which made the device more efficient. And he would work these out in consultation with Eickemeyer as he proceeded with the drawings.

Nothing is on record to show that Steinmetz improved any of Eickemeyer's inventions in this way, but it is quite possible that he suggested modifications which, being satisfactory to Eickemeyer, were incorporated into the finished design with his full approval.

Eventually three electric street-cars were equipped with the motors designed by Eickemeyer and Field, their construction having been carried out from the drawings of Steinmetz and the other draftsman. These three cars were first operated, in trial runs, on the Steinway Road, in Brooklyn.

Almost without exception, the trials were made on Sundays. Every Sunday morning, a party of hopeful men eagerly set out from Eickemeyer's factory, Steinmetz among them, to see the new electric cars start on their experimental trips.

Passengers were not lacking, for it was a new thrill, with all the elements of a novel adventure. The public, however, treated the electric cars pretty much as a joke. People rode in them because it was a unique "stunt" to enjoy the sensation of traveling over the ground without being drawn by horses. Nevertheless, for the practical business of getting anywhere at any specified time, they still relied on the faithful horse-cars or their own two legs.

It must be confessed that this was hardly more than natural, since scarcely a week passed that one or more of the electric cars did not break down. Usually the car that started in the morning had to be ignominiously towed back to the factory, more or less completely crippled. Another would then be sent out to take its place. Sometimes that one, too, went wrong. Then the third and last would be called upon, and that occasionally ran until afternoon. The week that followed was spent in repairing the cars for a repetition of the trials on the succeeding Sunday.

The Eickemeyer-Field cars presented one especially vexatious problem. The trolley-wheel at the end of the pole, forming the contact-point by which the electric current traveled from the feed-wire to the motor of the car, could not be kept in place on the wire. As the first cars constructed were provided with two trolleys, the wheels of which ran on top of the wire, instead of beneath it, continual watchfulness was necessary to keep them where they belonged.

Field once conceived the plan of making a magnet out of the trolley-wheel. His idea was that by magnetizing the wheel it would always attract the wire to it, so that the two would be continually glued together, so to speak, yet without preventing the wheel from traveling over the wire as the car proceeded.

The plan seemed so logical that it was decided to give it a trial. But the actual test proved disastrous. The moment the car started, the heavy wheel pulled off, and down came the whole trolley-pole, almost crashing through the roof of the car. That wound up the experiment with magnetized trolley-wheels in a pretty conclusive manner.

This was the period in which Eickemeyer brought out an electric motor especially designed to operate elevators, at the request of Norton P. Otis, an old friend, of Otis Brothers & Company. It was a shunt-wound motor, one of the first of its type to be developed. Eickemeyer then invented all of the required devices for its control by the operator in the elevator-car.

The outcome of this was the organization of the Otis Electric Company, a branch of Otis Brothers & Company, which built the first successful commercial elevator. It soon developed into a flourishing business which expanded into the big industry that it is to-day. When the Eickemeyer plant was bought up by the General Electric Company, the business of manufacturing the elevator-motors was not included in the sale but was taken over bodily by the newly formed Otis Elevator Company.

All these experiments were intensely interesting to the inquiring and observing young draftsman, who was working out most of the designs for the experimenters. Nothing was lost upon Steinmetz, but on the contrary he quietly absorbed the succession of ideas which these events brought before him. He realized more or less clearly that he was watching the never-tiring, never-defeated genius of human invention, which was bound to win in the long run and would stick to it until the victory was realized.

But Steinmetz, for the most part, kept himself in the background, for these men were older and more experienced than he was. They were the inventors and master minds of the moment; he was, just then, an unknown worker at desk and drawing-board, busy with his T-square, compass and rule, although he could discard his mathematical tools much more safely than could most designing draftsmen, on account of his natural aptitude for mental calculations.

Then one day, as will appear in the pages immediately following, his daily work, in relation to these electric motors for trolley-cars, caused him to do some important research work of his own, out of which proceeded his first astonishing discoveries and mathematical formulæ, establishing new and most valuable practices in the whole broad field of electric motor design.

CHAPTER XI

EMPLOYER and employee, in the case of Eickemeyer and Steinmetz, gradually came to have relations that were exceedingly warm. Eickemeyer, as the presiding genius of the whole factory organization, kept a keen oversight of all the work done in the plant. Every day he went on tours through the factory, observing what was in progress in each department.

The workers, both high and low, looked up to him as a remarkable man and one who, like themselves, had mastered the machinist's trade. He was tall, straight as an arrow, with the head of a patriarch, and they felt a certain degree of awe in view of his brilliant inventive faculty. But in all daily working relations, the entire place was like a big family. Eickemeyer was a true exponent of thoroughness, exacting, with an inborn dignity, but by no means severe, and he always recognized, encouraged, and enjoyed the slightest trace of originality in his workers.

It was inevitable, under these conditions, that Steinmetz should attract his attention. This first happened about two weeks after Steinmetz began his duties, when the two met in a mutually agreeable consultation. It came about somewhat peculiarly, during one of Eickemeyer's daily tours through the plant. Eickemeyer entered the drafting-room abruptly, to inquire: "Does any one know what will take an aniline stain off my fingers?"

"Yes," promptly answered Steinmetz, "sulphuric acid will do it."

And then he explained how to use the sulphuric acid to obtain satisfactory results. Eickemeyer stayed several minutes, talking with Steinmetz about his work, and observing his enthusiasm. He found Steinmetz so alert and so appreciative of research work that from that time forth the drafting-room became one of his favorite stopping-places. Some months later, as will be seen, this eager interest of Steinmetz in the work he was doing led to his selection as the logical man to establish an experimental laboratory on a small scale.

Socially the two were equally congenial. The brilliant, hard-working young draftsman was able to make himself at home almost anywhere; and before long the "big boss" invited him to make a Sunday afternoon call at the Eickemeyer home, Seven Oaks. The family was cordial to the young stranger, who fitted into the home circle most congenially.

Soon it was the practice for Steinmetz to call every Sunday, spending most of the time between three and ten o'clock there. Other visitors also came, for Eickemeyer liked to keep a semblance of

FIRST CLAIM TO RENOWN

open house Sunday afternoons and evenings. A mingled scientific and social atmosphere had developed through the years, which appealed very strongly to Steinmetz's ever-gregarious leanings.

The time was spent, as Steinmetz described it, "in discussing what had been done during the preceding week with the various lines of investigation and experiments in which we were both interested. We would talk about these matters all the afternoon. Then we would have supper, and after supper we would meet friends of Mr. Eickemeyer's and have social visits together."

Martin refers to these Sunday afternoon visits in discussing the fraternal intimacy that grew up between Eickemeyer and Steinmetz, an association much closer than existed between Eickemeyer and any other man in the establishment. If Eickemeyer had been out of the office for a period, because of ill health, he would go through the whole plant upon his return, to see how things were going.

"Finally," writes Martin, describing such a trip of inspection, "came the testing laboratory and the man in charge [Steinmetz]. The relations here were quite different. Eickemeyer would sit down and talk with Steinmetz by the hour. When we remember that, whatever their actual experience in the field of applied electricity, here were two of the most capable minds which that science has yet known, it may be judged that these were illuminating conversations; and the present-day engineer would be prepared to give much for a record of them, if such had been preserved.

"These laboratory consultations were conducted probably with no more formality than was observed when the men met outside the factory. Steinmetz and Eickemeyer were a great deal together, and both had this matter of the magnetic qualities of iron greatly at heart. The subject, it may readily be imagined, was therefore constantly under discussion; and the speculation occasioned by some result which may have been found during the week's testing was allowed full scope in the meeting on the following Sunday afternoon, which Steinmetz soon fell into the way of habitually spending at the big Eickemeyer house on the hill. He was living meanwhile with Edward Mueller, his predecessor in the drafting-room, and at this time the senior draftsman to the factory."

Those Sunday evening gatherings at Seven Oaks were rich in conversational repartee, vivacious with social good feeling and constantly tinged with moments of historical interest. Apparently the company frequently numbered a dozen or more; and when the supper-hour arrived, the big table in the dining-room was made still bigger, until it stretched almost from one wall to the other, leaving scarcely room enough for the servants to squeeze through as they passed to and fro in waiting upon the guests.

At the head of the table was Eickemeyer, a commanding figure with alert countenance, keen, quick, discerning eyes, and full flowing beard, looking the part he played in the life of the city as the patriarch of Yonkers. The visitors were people from various professional and occupational classes, men with technical leanings appearing somewhat in the majority.

A personage frequently seen at the table was Robert Parkhill Getty, one of the real pioneers of the town, whose boyhood had been lived in New York when the present metropolis was little more than an overgrown village. He had been part and parcel of Yonkers since his young manhood; consequently his personality was strongly stamped upon the community.

In 1857 he had been a village trustee, and he was intermittently president of the board until 1869. With the attainment of a city charter, he became water commissioner, and in 1881 he entered upon nine years' service as city treasurer.

Mr. Getty was one of the men whom Steinmetz thoroughly enjoyed, for the old pioneer had a delightful fund of anecdotes, which went back to the provincial days of New York. How quaintly rustic it must have seemed to hear him relate the scenes he witnessed around his boyhood home, which faced a little square in down-town New York, where village customs still clung with winsome tenacity, and where there was a village pump to which he often saw Mrs. Commodore Vanderbilt come down with her pail to draw water with her own hands like the good housewife that she was.

CHARLES PROTEUS STEINMETZ

Those were the days, too, when the entire area north of the old New York City Hall was nothing but a series of grazing fields for cows, with a few settlements scattered here and there; and when the City Hall itself was "two-faced"—finished handsomely in front, but plain, unadorned brick at the back, on the theory that nobody would have occasion to go beyond that point, as the city all lay in the other direction. If a person wanted to go up to what was later 125th Street, Mr. Getty recalled, it was necessary to hire a horse and carriage and start the day before!

The heartening circle around the long Eickemeyer table sat in charmed silence as a rule, Steinmetz with the rest, when Mr. Getty loosed the flood of his recollections.

And there was Steinmetz himself, perfectly at home amid the interchange of wit and wisdom. An eager listener, a quick, enthusiastic talker, conversing not alone on technical matters but on many other subjects that chanced to receive a hearing, he was a noticeable companion on these pleasant occasions.

Steinmetz was alike admired and respected by those who met him at Eickemeyer's Sunday night assemblies and who came to realize the character of the man. The glances that turned in the direction of this odd little person who was barely visible above the table-top, were entirely friendly, never disparaging, although his actions were peculiar to himself. He had a way of sitting poised on his chair so that he could feel capable of passing a dish or taking whatever was handed to him.

He would generally adopt a method wholly original in securing a biscuit or a slice of bread without the trouble of having them passed to him. With feet resting on the rungs of his chair, he would balance himself skilfully and, when properly poised, would suddenly lunge forward, fork uplifted like a harpoon, to spear a biscuit, pause a moment to maintain his balance, and then drop suddenly backward into his seat. In all this manœuvering, let it be said, he was wholly unconscious of any unconventional effect; it was merely his own naïve manner of meeting the immediate situation without causing inconvenience to others.

On one of these evenings, a young member of the circle, from motives best known to himself, suddenly said, "Please, Mr. Eickemeyer, what is the difference between the yolk and the white of an egg?"

"Ask Steinmetz to tell you," exclaimed Eickemeyer, with a smile, as if he knew what was forthcoming.

Whereupon, to the surprise and delight of the questioner, Steinmetz entered upon a dissertation as to the chemical and other qualities of an egg. He was so conversant with the subject that he spoke as if the inquiry had unlocked a perfect reservoir of stored-up knowledge; and he held the earnest attention of the entire group for fifteen or twenty minutes.

"It seemed," said the younger Eickemeyer, many

years later, "as if words of explanation from Steinmetz were, like a great teacher, ever on the tip of his tongue, only awaiting the formal question to burst forth so readily as to be like water from a fountain. For his was a never-ending fountain of knowledge; and, to one intellectually hungry, sitting at table with Steinmetz was to be the guest at a banquet."

In truth, Steinmetz usually commanded close attention whenever he had anything to say; and he and Eickemeyer were absorbed for hours in technical discussions during the long Sunday afternoons, to the wonderment of all who chanced to hear them. They nearly always talked on electrical subjects during these Sunday visits but once in a while switched off to more general matters. The younger Eickemeyer constantly found it fascinating to stroll into the library and listen to the talk of these two masters of research and experiment, both of them torches to their fellow-men, and one of them destined to become, not only a torch, but a blazing beacon upon a lofty hill.

Appreciating, as he did, a mind as original and alert as that of Steinmetz, Eickemeyer naturally came to discuss ways and means of expanding his investigations during the frequent talks which they had in the drafting-room at the factory. Within a few months he was led to speak of the possibility of a research laboratory for the plant. Such a project, he explained to Steinmetz, was much needed in view of the advanced point which he had thus far reached in his inventive work.

"Would you feel like establishing such a laboratory for us?" inquired Eickemeyer.

"Yes, yes!" answered the young draftsman. "I will do it."

And he did. He established it simply by taking possession of an extra room, acquired for the purpose by Eickemeyer, and going to work, with one assistant, of whose services he was not always assured. This was the Eickemeyer experimental laboratory; an unpretentious, rough little workshop, but one of the historic spots in the history of electrical science, for it was here that Steinmetz did the work which led to his revolutionary discoveries in the field of magnetic losses, or the law of hysteresis, as electrical engineers term it.

"It was probably about March of 1890," says Martin, in the "General Electric Review," "that he began to specialize on magnetic testing. For the purpose of carrying out this work on a large scale and in a more thorough manner, Eickemeyer leased a room adjoining his existing factory, in which the testing equipment became installed.

"The laboratory itself was about 20 feet wide by 30 feet long, and at the present time [1912] shows little signs of its use twenty years ago as a magnetic research room. On a recent visit to Yonkers, the only piece of equipment which Dr. Steinmetz could trace as having held its place unchanged since his time was a very rusty wash-stand in a corner of the room by the window. The laboratory (as it then was) is located on Nepperhan Street, Yonkers, and the passenger to New York can easily see the building to-day as his train passes through Yonkers station on the New York Central line. Although the Eickemeyer Company was bought up at the end of 1892, the name 'Eickemeyer & Osterheld Manufacturing Company' is still standing on the north wall.

"Steinmetz was placed in charge of the new laboratory. His assistant, a man skilled in the hat machinery business, was liable to be called out at a moment's notice to investigate and remedy troubles in the machines used in hat-making."

Dr. Steinmetz recalled the days in this experimental laboratory with pleasant zest and modest reference to his own activities therein.

"The laboratory," he said, "came about as a matter of course. When I first began to work at Eickemeyer's, there was one other draftsman there, Mueller, at whose home I finally boarded. We were in the same room together.

"Later the second floor of the next building to Eickemeyer's factory was secured, a door was cut through, and that became my own room. That was where I worked from that time forward, where the laboratory was set up, and where I made my investigations into the law of hysteresis. At first I had a shopman working under me, and later a draftsman.

"The shopman did the mechanical work of the lab-

A humble shrine of American electrical engineering the laboratory of the Eickemeyer establishment where Steinmetz made his hysteresis studies. The window on second floor of middle building, directly above the side door, indicates the room used as a laboratory

oratory. He was not there all the time, as he was an expert repairer of hat machines, and if anything went wrong with these machines he was sent for immediately. Thus he would disappear from my room for a day or two days at a stretch; but I had him most of the time, even so. This was Eickemeyer's research laboratory—that one man and myself. Later, as I have said, a draftsman worked under me.

"This was the room where Stephen D. Field used to come and sit by the fire. It was a cold room in the winter; so I bought a good-sized stove. It took a big fire to keep the stove going, and that made the room roasting hot. When I tried to check the fire, it would go out.

"We used paraffin paper to revive the fire, and later from the pattern-shop we procured linseed-oil, which was rather explosive if used too freely. On one occasion I blew off the top of the stove with it."

The laboratory soon developed into the most jovial spot in the factory. The men with Steinmetz were fond of a joke now and then, and Steinmetz himself relished the pranks which occurred there. In one of these the good-natured Field was the victim.

Field, it appears, enjoyed lounging in the laboratory and talking. He had a favorite nook beside the stove; and one day, while he was dozing there, one of the men poured a cup of linseed-oil on the fire, ostensibly to rekindle the blaze. The oil, used so generously, blazed up with a roar and a mild explosion. Field was awakened with a jump and started out of his chair in momentary panic. The onlookers, meanwhile, were laughing in suppressed mirth, Steinmetz among them, as they observed Field looking sheepishly about to see who had disturbed his slumbers. After that he never sat down by the fire without first glancing about to see if any one was near enough to play tricks.

Some time later a new-comer appeared. This was Tischendorfer, a young chap just arrived from Germany, and believed to be the draftsman to whom Steinmetz referred. The latter described Tischendorfer as "the Dutchest Dutchman I ever saw"; and it was decreed that he should undergo an initiation. Linseed-oil was again utilized, this time for the purpose of dipping a cigar.

The cigar was presented to Tischendorfer, who lighted it and began smoking. As the cigar burned down, the linseed-oil gave off an almost unbearable odor, which soon filled the whole room. Tischendorfer, however, turned the joke on the jokesters by smoking the cigar down to the smallest butt with apparent unconcern, while every one else was made extremely uncomfortable by the atmosphere that was created.

"It was about this time," comments Martin, "that the want of a more thorough understanding of the magnetic properties of iron was felt by alternating current designers very acutely."

There were two engineering text-books upon this

general subject, which were consulted at that period by all designers of electrical apparatus. These were the published tables of Ewing, "which left off at the very point where a designer of a commercial machine might have found them useful"; and the theory of the magnetic circuit, which was clearly understood and had been treated in a book by Gisbert Kapp, published in 1886.

The whole big problem, as it cropped up persistently before the electrical engineers of that day, is not difficult to grasp with an elementary, non-technical explanation; and we have just such an explanation, by none other than Steinmetz himself, who was amazingly skilful in the rôle of teacher, able to explain with clearness a subject which appears intricate to understand. And here is what Steinmetz says of the dilemma facing electrical pioneers in the last decade of the nineteenth century:

"In most electrical apparatus magnetism is used. Sometimes the magnetism remains constant, as in the fields of direct current machines; sometimes the magnetism alternates, as in transformers.

"When the magnetism alternates, it consumes power. Such power consumption means loss of efficiency and results in heating. It is therefore of importance to the builder of electrical apparatus to make the designs so that this loss of power by alternating magnetism (called 'hysteresis') is as small as possible.

"However, the laws of this power loss were entirely unknown at that time, and many engineers even doubted its existence. The designer of electrical apparatus simply built the apparatus, then tested them, and when the hysteresis loss was found too high and the efficiency too low, or the machine too hot, they tried again. This, obviously, was not a satisfactory way.

"Now, in this experimental work I was Mr. Eickemeyer's assistant, and I had to calculate and design an alternating current commutator motor, of the same type as that now used on the New York, New Haven, and Hartford Railroad and other railroads. I knew there would be a loss of power in the alternating magnetism of the motor, and I wished to calculate this 'hysteresis loss,' to get the efficiency of the motor. I therefore looked through the literature obtainable and found two tables of hysteresis losses given, one by Ewing, in his book on magnetism, and one by Kapp, in his little book on alternating currents.

"Unfortunately, the two tables disagreed with each other very much, and the curves given by the tables differed in shape from each other. I then studied both tables and found that Kapp's table must contain a typographical error.

"From Ewing's table of hysteresis losses, however, I derived mathematically a law, the 'law of hysteresis,' showing how the hysteresis loss increases with the increase of magnetization. Roughly, it is that every time the magnetization doubles, the hysteresis loss trebles. This law of hysteresis, as derived from Ewing's data, I published in the 'Electrical Engineer' for December 9, 1891. Then I started testing myself the various kinds of iron and steel, and other magnetic materials which were available, and gave the results of these tests in the first paper on the law of hysteresis, before the American Institute of Electrical Engineers."

That, in the simple language of this master mathematician, is the story of his discovery of the law of hysteresis losses. To the uninitiated it may appear to be merely a somewhat expanded process of routine, which does not sound particularly notable when so easily stated. But it was a process which no electrical engineer of the period had been able to work out; a process that cleared up, for all time, a problem that was vexing every designer of electrical machines.

As one scientific authority has expressed it: "In plain words, he gave a method by which engineers can figure how much magnetizing current they should use to magnetize a given piece of iron to be used in an electrical generator or motor, to throw off so many lines of magnetic flux (or flowing of a magnetic current). They could thus figure how many watts of loss there would be in the iron, and therefore how hot the iron would become when worked in given conditions."

The significance of it all is pointedly indicated in Martin's account in the "General Electric Review." He thus describes the occasion for seeking this law, and how Steinmetz, with his intense mathematical mind, penetrated to the great secret:

"Steinmetz and Eickemeyer were now [late summer of 1890] engaged on the design of the single-phase commutator motor with compensated winding [Eickemeyer's best-known contribution to the electric storehouse]. Both men realized the limitations placed upon them by this 'groping in the dark.' Steinmetz took all of Ewing's results which he could lay hands on and subjected them to a very critical examination. He probably suspected that he would . . . discover 'that law of nature which gives the dependence of the hysteresis upon the magnetization.'"

Martin quotes Steinmetz, at this point, as saying: "The results of my calculation seemed to me interesting enough to publish, in so far as all these observations fit very closely the calculated curve within the errors of observation; and the exponent of the power was so very nearly 1.6 that I could substitute 1.6 for it."

Continuing his comments, Martin says: "Great and important as is the work which Dr. Steinmetz accomplished in later years, there is a point of view from which it may be said that his chief claim to greatness rests upon the establishment of this fundamental law of magnetism, a law which, it is true, was announced empirically, but which may at some time in the future be found to possess a real physical significance.

"This investigation took Steinmetz through the

summer and fall of 1890; and by the time his article appeared, he was already at work on hysteresis tests in the Yonkers laboratory, much more complete than any which had been made up to that time, carried out on any and every sample of iron which could be obtained, and employing what seems to us the crudest power plant and testing outfit. They were nevertheless to be massed into the paper before the Institute which was to establish his growing reputation upon a permanent and unassailable basis."

This famous paper, which was really two papers, or one paper divided into two parts, created a sensation among electrical engineers, as well it might. The first part was read before the American Institute of Electrical Engineers on January 19, 1892, barely more than five weeks after the appearance of the article in the "Electrical Engineer." The second part followed nine months later, on Stepember 27, 1892.

Approximately two hundred pages in the proceedings of the Institute are taken up by these papers. The actual length of the first was forty-eight pages, while the second took up 130. This is a total of 178; but in addition Steinmetz contributed at considerable length to the discussion that took place after the presentation of the papers.

Nobody who is not a member of the electrical engineering profession, or endowed with a love for mathematics, can appreciate, from what has thus far been said, how deeply mathematical these papers were. Least of all can they realize this from anything that has been quoted from Steinmetz himself.

Yet even to the keenest mathematical minds in the assemblage of engineers at the Institute's meetings they were extremely impressive, not alone for the invaluable formulæ which they placed at the disposal of the profession for the first time, but fully as much for the brilliant, intensive process of calculating which they revealed. The technical editor of the "Boston Transcript" summed it up adequately when he recently wrote:

"Steinmetz made himself famous, so far as electrical men were concerned, by a masterly paper before the American Institute of Electrical Engineers in 1892. That paper dealt with a subject upon which nearly every one at the time was pretty meagerly informed. It was not written in brilliant English, and many of its 200 pages were filled with mathematical signs and an advanced reasoning, which put the whole work on a plane considerably above the head of the average electrical engineer of the day. But from it all emerged a fact and a law—a fundamental law of magnetism. Steinmetz found an exponent connecting the magnetization with the hysteresis loss. . . .

"That alone should make a man famous; because it was nearly all pure pioneering; was based on work which had to be carried out with the crudest possible power plant and measuring outfit; and because it furnished results which were of immediate commercial

Steinmetz in 1890, just before he read his famous paper on the law of hysteresis before the American Institute of Electrical Engineers

importance. That investigation into the phenomena of the magnetic circuit, and the discovery of the 1.6 law, was one of the most valuable things which Steinmetz has ever done."

As remarkable as any characteristic of these astonishing expositions by this young man of twenty-seven, with his sadly crippled body but his powerful brain, was their completeness. In his second paper, to cite one example of this, Steinmetz included comprehensive data on the magnetic constants for all magnetic materials known at that time. He also gave equally thorough data on the magnetic characteristics and the saturation values, so-called, of these materials. That is, he showed the greatest value of magnetism which a given material could carry.

He showed that cast-iron, extensively used at that time, will carry only about half as much magnetism as wrought-iron; and that cast-steel, then just coming into use in electrical machinery, is intermediate between cast-iron and wrought-iron. Some grades of cast-steel, he revealed, will carry magnetism in values as high as wrought-iron.

It has been pointed out that the discovery of the law of hysteresis brought Steinmetz into acclaim with extraordinary swiftness, especially so when it is realized that he had but three years previously landed at Castle Garden as a steerage passenger who could not speak enough English to impress the immigration officials. But it is worthy of note that any engineer who was then designing electrical machines was confronted with difficulties in dealing with the magnetic circuit, difficulties which seemed perfectly insurmountable in the light of any working data obtainable. He was therefore very ready to give a hearing to any theory that appeared to show the way out. Furthermore, there are indications that during 1891, the year in which the experiments at Yonkers were conducted, many of Eickemeyer's friends who were prominently placed in the electrical profession visited the laboratory and became more or less familiar with what was going on there.

The field was thus unquestionably fertile for the potent seed which Steinmetz sowed therein when he stood before the gathered celebrities of the American Institute of Electrical Engineers and sprinkled mathematical equations among them with a precise, and withal a liberal, hand.

All men identified with the electrical profession now knew that a new thinker, a powerful analytical mind, had arisen among them in the person of this young man of dwarfish stature with the quick manner and the gentle, friendly eyes. When President Frank Sprague introduced Steinmetz for the second section of his paper, he remarked upon the reputation already obtained by the author in this branch of research, adding, "His work in the past has been most important in its character, and this paper will fully support the reputation he has already earned."

Nevertheless, the discussion which followed the

paper suggests that the new theory, with its very complete verification by the author, was just a little over the heads of the electrical authorities of the time. Those who were fully capable of appraising it accorded it instant recognition; and the rank and file followed inevitably in due course of time.

Of the several comments expressed during the meeting of September 27, 1892, that of Dr. A. E. Kennelly is noteworthy. "I think it will be unnecessary," he said, "for me to express the general and very high opinion in which we hold the paper to which we have just listened. It is a classic to us, and I think it will be a classic to a great many more than ourselves. The Institute may well congratulate itself upon this paper."

Mr. William Stanley, one of the pioneers in his own field of electrical research, remarked: "It seems to me that Mr. Steinmetz has done for the magnetic circuit very much what Ohm did for the electric circuit. He has defined the law relating loss of energy to flux. To the constructing engineer, working with the alternating current appliances of to-day, the paper affords more assistance than anything we have ever listened to."

Applause was also granted by Dr. Charles E. Emery when he said: "This paper has evidently required an enormous amount of earnest work. It is a very notable example of successful experimental investigation, for which, as well as for the clear and complete manner in which the subject has been examined and presented, the author is indeed to be congratulated."

The attitude of the profession is finally summarized by the expressions of electrical magazines of the period, especially the "Electrical Engineer" of January 27, 1892, and the same publication for October 19, in the same year. In the former issue, speaking of Steinmetz's first paper, the statement is made that "the American Institute of Electrical Engineers has been the medium for bringing out not a few papers of scientific interest and practical importance; but we believe that none of more absorbing interest and practical utility has been presented to the Institute than that of Mr. Charles P. Steinmetz last week on the law of hysteresis."

When the second section of the paper was presented, the "Electrical Engineer," on the latter date mentioned, said that "he thus finally brings the magnetic circuit within the reach of analytical treatment, in the same way that the electrical circuit has been mastered by Ohm's law. . . . He holds out to us the hope of our obtaining a full understanding of the phenomenon of magnetism in the near future."

The stir occasioned by these papers and the attention which they centered upon the person responsible for them did not dim in Steinmetz's mind the assistance which his association with Eickemeyer had given him. Both then and ever after he credited Eickemeyer and the latter's magnetic bridge with helping him along the path to final success.

"Mr. Eickemeyer had designed and built," he says in his reminiscences, "a very ingenious instrument, the magnetic bridge, which permitted the comparison of the magnetic carrying capacity of different materials.

"He adopted a sample of very soft Swedish iron as standard, and gave the quality of the other iron samples in percentages of this standard. Thus, in building electric machines, when casting the field frame, a piece was cast with it, cut off, turned to size and magnetically tested, the field winding of the machine being then calculated from this data. Then we made extensive tests on cast steel, for the various manufacturers of cast steel.

"Mr. Eickemeyer, being very much interested in magnetism, encouraged my investigations, and for a long period I spent practically all my time in magnetic research and testing. Mr. Eickemeyer materially assisted with advice and kindness, so that considerable credit for the results of the magnetic investigations is due to him.

"Much of the work on the determination of the constants given in the second paper was done on Mr. Eickemeyer's magnetic bridge. While this instrument had the disadvantage of giving only comparative results, when carefully handled it was very satisfactory, and its use constituted probably the first systematic testing of all the magnetic materials used in electrical machine manufacture."

The effect of the law laid down by Steinmetz became manifest as time went on. Electrical manufacturing processes were stimulated, and electrical engineering was aided and enlightened. The induction motor and the polyphase motor, as now known, were perfected by utilizing the new formula.

And the reaction toward the personal status of young Steinmetz was pronounced. He became widely known among electrical scientists. His mathematical genius was recognized as far above the ordinary.

Nevertheless, this young man, not yet thirty, continued quietly his daily work in Eickemeyer's establishment at Yonkers, unperturbed by the sensation he had awakened; unperturbed, also, even in the face of events which began to shape themselves in this year that were to precipitate another change in his fortunes and his physical environment.

CHAPTER XII

THE ALTERNATING CURRENT MEETS ITS MASTER

STEINMETZ in his later years once made a modest estimate of his life's work. It was as near as he ever came to writing his memoirs. In a few very sketchy pages he reviewed the big things of his career, as he saw them in the retrospect.

He disclosed the interesting circumstance that, viewing his life as a whole, there were three "most important" achievements. And first of these three to be named by him was his investigation of magnetism. The second was his development of a practical method of making calculations with the baffling alternating current. The third was his general study and theory of "electrical transients." He never actually completed this last, for it embraces the pranks and upheavals played upon electrical systems by lightning, in itself a broad field of research, and with which he was still busy when death suddenly closed his career with a dramatic swiftness equal to one of those very lightning-flashes that he was so fond of theorizing over.

But our immediate concern, at the period in his career now under consideration, is with the second of that mighty trio of mathematical masterpieces by which the whole profession of electrical engineering has benefited.

All the evidence at hand would seem to indicate that Steinmetz passed immediately from his work on the law of hysteresis to his study of the alternating current problem. The one rather naturally opened into the other, since the discovery of the law of hysteresis brought on a greater field of usefulness for machines using alternating current, as distinguished from direct current.

Steinmetz undoubtedly perceived this himself, or, what is more probable, he foresaw it. Hence, no sooner had he completed his intricate mathematical work upon the magnetic theory than he embarked upon an even more intricate line of mathematical research in seeking to solve the puzzle of the alternating current.

It is hardly possible to exaggerate the great service to all enlightened peoples which Steinmetz performed in both of these investigations. The electrical profession benefited most directly, to be sure. But this is the electrical age in the world's history; not a day passes but that nine tenths of all Americans either avail themselves directly of some service of electricity or are served by this mighty force indirectly. Every time they ride on a street-car, every time they take an elevator, every time a housewife's tired arms are spared by the willing motor in her electric washing-machine or some essential industrial process moves smoothly forward under the drive of a motor many times larger, it is a recognition that electricity is the burden-bearer of the race.

And these motors, whatever their size, if they use alternating current in their operation, are every one of them a purring, whirring, power-producing monument to the mathematical genius of Steinmetz. For every one of them has embodied in it, as its indispensable foundation of efficiency, that fundamental law of hysteresis losses which Steinmetz successfully worked out.

Moreover, since it is probably true that alternating current serves the every-day rank and file of the population much more frequently than does direct current, Steinmetz further made his life of undying usefulness to all future generations by providing electrical engineers with a method of making their alternating current calculations. That achievement enabled these electrical engineers from thenceforth to develop to the utmost limit and with the utmost celerity the possibilities of electrical application.

In electrical engineering, alternating current is one of the two kinds of electrical current in every-day practical use. The other kind is direct current.

The difference between the two is very distinct. Direct current flows continually in the same direction. Thus it can be measured in amperes (the ampere being the unit for measuring the flow of an electric current), and its action can be calculated numerically in a simple manner.

But alternating current does not flow continuously in the same direction. As its name implies, it alternates, flowing first in one direction, then reversing and flowing in the opposite direction, then reversing again and flowing in the original direction, and so back and forth, back and forth, usually 120 times every second. In certain applications of electricity, however, alternating currents are made to alternate not merely 120 times per second but thousands and thousands of times per second.

The alternating action was thus described by Dr. Steinmetz: "The current rises from zero to a maximum; then decreases again to nothing, reverses and rises to a maximum in the opposite direction; decreases to zero, again reverses and rises to a maximum in the first direction—and so on."

No wonder mathematicians were baffled by the peculiar nature of this mysterious force! And in dealing with it thirty years ago they were hampered by an imperfect understanding of the mathematical principles involved. Alternating current was particularly baffling in that, unlike direct current, it has no value and no direction. Its value continually changes, and so does its direction.

"Thus," says Dr. Steinmetz, in his explanation of this problem, "in all calculations with alternating current, instead of a simple mechanical value of the direct current theory, the investigator had to use a complicated function of time to represent the alternating current. The theory of alternating current apparatus thereby became so complicated that the investigator never got very far.

"True, practical electricians had been building and operating alternating current machines, and they secured, for practical use, a numerical value for the alternating current by means of an ammeter (an instrument for measuring in amperes the strength of an electric current). But you could not make any calculations with it."

Meanwhile the problem had been very much aggravated both by Steinmetz's own work on the law of magnetism, as already shown, and by the invention of the alternating current transformer. This type of electrical machine is one of the essential units in long-distance transmission of electric currents.

By means of transformers, electric currents can be sent for miles and miles from a distant generating station, at a relatively high voltage, and then reduced to a lower voltage upon arrival at substations, so that they can be eventually distributed to stores, offices, factories, and family dwellings at the customary 110 volts.

It was apparent to Steinmetz, as well as to every one else, upon the appearance of the alternating current transformer, that electric power transmission and distribution would thenceforth be by alternating current. Yet the alternating current theory was so hopelessly complex that the most astute engineer might have been pardoned for tearing his hair over it in despair.

Repeated efforts were made, before Steinmetz's research, to make use of various graphic methods; that is, by diagrams and geometrical figures. A whole series of diagrams for alternating current calculation had been worked out by Kapp, Blakesley, and others. But these endeavors failed to unravel the real tangle, which presented as insurmountable an obstacle as ever.

The dilemma, indeed, seemed utterly impossible of practical solution. Here was the alternating current, admittedly a most useful manifestation of electricity; a form of electric current which could be subjected to great increases or decreases in voltage by means of the newly invented transformer. Here was a means of sending useful energy to distant places, the very thing of which industry and society stood most in need.

And here, on the other hand, was the great barrier to its highest development and application, the fact that electrical engineers were wholly unable to find out just how far they could go in utilizing it, because they could not subject it to any satisfactory mathematical calculations.

The situation was tantalizing. The world of electrical men—and unconsciously the whole great world of affairs in general—was fairly yearning for a genius to appear with the magic key which would unlock, in a trice, the door to this thoroughly exasperating puzzle. And that is precisely what Steinmetz did.

As early as 1890 Steinmetz had begun to sense the challenge of the problem. As he became more and more absorbed in his studies of magnetism, the unruly character of the alternating current made itself increasingly apparent. For a while, before completing his tests under the new law of hysteresis losses, he was considering the nature of both these difficulties.

In his initial work on the subject of the alternating current, Steinmetz tried to do what almost every one else had attempted. He endeavored to apply the graphic method. He published a theory of the alternating current transformer in graphic treatment, this paper appearing in 1890 in the "Electrical Engineer," and in the same year being published in German in the "Elektrotechnische Zeitschrift," of Munich. This followed closely upon his studies of the same general subject while he was still in Zurich, where he published a theory of transformers (or "converters" as they were then called), without, however, even having seen one until after he came to America.

He quickly perceived the shortcomings of the graphic method. The impossibility of solving the problem by that means impressed him with inescapable finality.

"Such graphic methods," he said, in discussing the

subject, "have been used and are still used, to some extent, with success. They are good for getting a conception of the relation of the different alternating currents and voltages to each other. For calculation, however, they are of very limited value."

The final solution came through the application of pure mathematics. It involved a degree of intricate mathematical work that would bewilder the untechnical layman. In truth, the complete elucidation of the theory which eventually solved the puzzle, written by Steinmetz himself, for the enlightenment of the profession, fills three volumes. When he first expounded it as a paper before one of the scientific bodies, the introduction alone consumed far more than the time allotted the speaker on the program of the session.

For the simple purpose of following the thread of the man's career, and of molding an estimate of his genius, let it merely be stated here that Steinmetz found a mathematical method of reducing the alternating current theory to a basis of practical calculation; a fact easy to state, but tremendous to grasp.

Steinmetz has himself explained how he did it. As a preface to his explanation, which is interesting to follow even though it deals with processes beyond the grasp of those not mathematically inclined, it may be remarked that in higher mathematics there are symbols known as "ordinary numbers" and "general numbers," the former much less complex than the latter.

THE ALTERNATING CURRENT

"The idea suggested itself at length," to quote now from Steinmetz's statement, "of representing the alternating current by a single complex number, or 'general number,' as it is better called. This proved the solution of the alternating current calculation.

"It gave to the alternating current a single numerical value, just as to the direct current, instead of the complicated function of time of the previous theory; and thereby it made alternating current calculations practically as simple as direct current calculations. Indeed, the same calculations apply, except that the numerical value explaining the alternating current is a general number, while that explaining the direct current is an ordinary number.

"The introduction of the general number has eliminated the function of time from the alternating current theory, and has made the alternating current theory the simple algebra of the general number, just as the direct current theory leads to the simple algebra of the ordinary number."

So Steinmetz had now solved the perplexing problem. But to solve it was one thing; to make the solution clear, even to highly technical mathematical men, was quite another. The task of finding the way out had been accomplished. But a second task, as arduous as the first, still remained. This was the task of enabling others to see the pathway, of showing the blind the road leading to the light.

What happened next when Steinmetz, having worked out the new process, proceeded to expound it for the practical use of the electrical engineering profession, is naïvely related in his own words.

"When I realized," he says, "the enormous power of this method in resolving the apparently most complicated problems of alternating current theory into simple algebraic equations, I wrote out a rather complicated outline of this new method. I gave first the explanation of the method, and then its application to the more important alternating current circuits and apparatus.

"This paper I presented to the International Electrical Congress in 1893. The Congress was divided into three sections—theory, theory and practice, and practice. My paper was presented as the second on the program of a morning session in the section on 'Theory.'

"The first paper was a highly theoretical discussion, and to my dismay I watched one hearer after another silently rise, edge to the door, and disappear. I began to fear that I would have no hearers left.

"But fortunately, just before I began, a very interesting paper in another section, which had numerous hearers, concluded, and, being followed by a rather uninteresting address, caused most of the hearers in that section to leave. Many of them, feeling it their duty to attend the reading of more papers, came into my section, and so filled it up again.

"We had an hour allowed us for presenting each paper. If needed, we could get a ten-minute extension. I got two such extensions, each of ten minutes, after the hour allowed me was up; and by that time I was almost through with the introduction!"

Naturally, no one at the International Congress grasped the idea which Steinmetz had embodied in this formidable document, much less understood its significance. Steinmetz realized this but was hopeful that he would acquaint the scientific world with his theory when the paper was published. Somewhat to his dismay, he learned that the International Congress did not have sufficient funds to pay the cost of publishing his paper. The immediate result was to postpone for several years the time when the profession should become familiar with the procedure of alternating current calculations by the symbolic method, as this sort of equation is termed.

A curious situation faced Steinmetz during this period, a period which really stretches through all the years between 1893 and 1897, by which time he had become fairly well established in Schenectady as an employee of the General Electric Company. It was an experience that suggests something of the irony of genius, at least the irony that at times attends the career of a mathematical genius.

For in this interval of four years Steinmetz underwent the fate of the pioneer in applied mathematics. Having penetrated to realms hitherto untrodden, he found himself in unapproachable solitude.

He had brought forward a theory, a proposition;

and no one but himself was familiar with it, nor could any one else understand it! Steinmetz, with his symbolic method of alternating current calculation, was a perfect prototype of Einstein, with his theory of relativity.

The reality of this inability to grasp the new method becomes more emphatic in the light of the next series of papers which Steinmetz presented before the American Institute of Electrical Engineers. These papers went into the symbolic method more thoroughly, bringing out more of its practical side; for, as Steinmetz himself perceived, the symbolic method would remain unknown as long as his paper before the International Congress lay unpublished.

"I developed it [the symbolic method] further, in its application to all kinds of alternating current apparatus and phenomena," he says, "and presented it in a number of papers before the American Institute of Electrical Engineers.

"I believe it is due largely to the grand old man of the A.I.E.E., ex-Secretary R. W. Pope, that these papers were accepted, for I believe practically nobody read or understood them, as might be expected, since they used the symbolic method, which was still unpublished in the manuscripts of the International Congress papers."

Publication of all these papers finally came about through the efforts of Steinmetz himself. He negotiated with the predecessors of the McGraw-Hill Company of New York to bring out in book form not only the International Congress paper but also all the papers delivered before the American Institute of Electrical Engineers. This book presented the papers all worked up together into a general system, which became Dr. Steinmetz's first electrical text-book, under the title of "Theory and Calculation of Alternating Current Phenomena." It first appeared in 1897; and about this same time the proceedings of the International Congress, including Steinmetz's original paper on the symbolic method, were also finally published by the McGraw-Hill Book Company.

One would suppose that these publications would have been sufficient to introduce the symbolic method to electrical engineers generally. But Steinmetz perceived that it would be necessary to go further than this.

As the symbolic method was new to college instructors, and hence still newer to the students, he saw the need of a text-book which would explain this method satisfactorily for class-room use by students of electrical engineering. This was all the more essential since the general number, or complex quantity, although it was part of elementary algebra and introduced in the high schools, was so little used that students who later encountered it in alternating current studies had forgotten it and found it difficult to understand.

Hence, Steinmetz finally brought out another text-book, as an introduction to his work on "Alternating Current Phenomena," embracing not only an explanation of the general number as used in alternating current calculations but also a series of lectures given before graduate students of Union College. These lectures dealt with direct current apparatus, and with alternating current apparatus from a somewhat different point of view, and so suggested publication. The title of this new work was "Theoretical Elements of Electrical Engineering."

But the prolonged process of thoroughly acquainting engineers present and prospective with the new theory did not even stop here. Steinmetz wanted to get back to the very high-school period of explanation. He wanted boys who took a scientific course in high school, and who might incline toward electrical engineering in the technical college, to have the means of preparing themselves from the very beginning with the complex quantity and thus lay a solid foundation for their later work.

This led him to write "Engineering Mathematics," which he intended to be "a complete elementary treatise on the general number, the trigonometric series and the differential equation of electrical engineering as the three mathematical tools which are of fundamental importance in electrical engineering, and in which the mathematical preparation of the student is entirely inadequate, even in the college mathematics course."

While these various text-books were preparing, the

original work on "Alternating Current Phenomena" had expanded in bulk, through succeeding editions, until at length it became necessary to divide it into more than one volume. This brought about its publication in three volumes: "Alternating Current Phenomena," "Electric Circuits," and "Electrical Apparatus."

All this effort, with the one purpose in view of helping others to a clear understanding of methods which would be of immense value to them and of lasting benefit to society at large, betokens far more than the mere intensive mental activity of the keen-thinking scientist. It discloses also an almost infinite patience, combined with all the acumen of the teacher, the faculty for making clear to some one else what he understood himself.

And in all this there was further apparent a kindly, even a companionable attitude toward the many who at first were bewildered by the intricacies of the symbolic method. Here was none of the superiority of greater knowledge or the arrogance of unusual ability. Far from looking down upon his fellow-engineers because they did not readily grasp his theory at the meetings of the A.I.E.E., he himself gave the one obvious reason why they did not grasp it, the fact that his original explanation had been buried in some unpublished manuscripts, so that the new symbolic method was naturally unknown. A courteous way of saying that of course if they knew the working of the symbolic method, which most of them had forgotten with the years, they would easily have comprehended its application to alternating current calculations!

Steinmetz received his reward, if he regarded it as a reward to see his theory stand the test of practical usage, after it came to be understood; and finally to see this method, which at first had bewildered all who tried to use it, become so generally adopted and so successfully used that even the most complicated phenomena of alternating current distribution, such as are dealt with in the great transmission-lines of to-day, no longer have any terrors for those who work with them.

"They are now easily calculated," modestly re-marked Dr. Steinmetz, many years later. "And with the great development of alternating current engineering, in power transmission and distribution, we never have any phenomena now which cannot be solved and calculated."

Yet, as already observed, this is due fully as much to the ability of Steinmetz to play the teacher to a whole rising generation of electrical engineers as to his genius in originally solving the tangle. He wrote a different type of book for every one concerned, clear back to the boy in high school.

He was, in a very literal sense, the great torch lighting the way into new realms; and without his guiding beacon-light, his reassuring genius in finding the way, his unwavering foresight as to whither the unfamiliar path would lead, America's electrical development would have been but a stumbling and a groping progress.

Long before Steinmetz had witnessed the universal adoption of his symbolic method, indeed, only a few months after his final paper on the law of hysteresis losses, the Eickemeyer & Osterheld Company passed out of existence as a separate business establishment. The interests of the company were purchased by a recently formed concern, which had come into existence on April 12, 1892, the General Electric Company.

When this new business organization was established, by the union of the Edison General Electric Company, of New York, and the Thomson-Houston Electric Company, of Lynn, Massachusetts, its immediate policy was to buy up any electrical manufacturers, with their patents, which looked promising.

The suggestion of Eickemeyer's establishment as one of these was first made by Norton P. Otis, of Yonkers, one of the founders of the Otis Elevator Company. He expressed the opinion to the officials of the new General Electric Company that Eickemeyer had inventions and manufactured products which the merged company could not afford to ignore if it wanted to make a good start toward electrical development in general. Negotiations followed, culminating with a meeting of the interested parties in New York, and the arrangement, at that time, of satisfactory terms.

CHARLES PROTEUS STEINMETZ

Primarily, the General Electric Company bought Eickemeyer's business because of the latter's patents and valuable electrical applications. Secondarily, it is quite evident that the General Electric Company wanted the services of the youthful mathematical master, Steinmetz.

The news of the whole transaction was imparted to Steinmetz by Eickemeyer, who told him that it had been arranged for him to be transferred to the General Electric Company. Steinmetz appears to have consented without the slightest hesitation.

A few days later, Steinmetz met some of the General Electric officials, who came to Yonkers to see what they had bought and to confirm the understanding that Eickemeyer's brainy engineer was to enter the new organization. At different times during this period Steinmetz met for the first time E. W. Rice, Jr., later president of the General Electric Company and now honorary chairman of its board of directors; G. E. Emmons, who eventually became manager of the Schenectady works and is now vice-president in charge of manufacturing; and Dr. Louis Bell, in charge of new developments for the company.

Mr. Rice in particular was the representative of the General Electric Company who interviewed Steinmetz on the question of entering the employ of the General Electric. He describes his visit to Eickemeyer's plant, and his first meeting with young Steinmetz, as follows:

THE ALTERNATING CURRENT

"I was then in charge of the manufacturing and engineering of our company, and my views were sought as to the desirability of acquiring Eickemeyer's work. I remember giving hearty approval, with the understanding that we should thereby secure the services for our company of a young engineer named Steinmetz. I had read articles by him which impressed me with his originality and intellectual power, and believed that he would prove a valuable addition to our engineering force.

"I shall never forget our first meeting at Eickemeyer's workshop in Yonkers. I was startled, and somewhat disappointed, by the strange sight of a small, frail body, surmounted by a large head, with long hair hanging to the shoulders, clothed in an old cardigan jacket, cigar in mouth, sitting closs-legged on a laboratory work table.

"My disappointment was but momentary, and completely disappeared the moment he began to talk. I instantly felt the strange power of his piercing but kindly eyes, and as he continued, his enthusiasm, his earnestness, his clear conceptions and marvelous grasp of engineering problems convinced me that we had indeed made a great find. It needed no prophetic insight to realize that here was a great man, who spoke with the authority of accurate and profound knowledge, and one who, if given the opportunity, was destined to render great service to our industry.

CHARLES PROTEUS STEINMETZ

"I was delighted when, without a moment's hesitation, he accepted my suggestion that he come with us."

As revealed by Mr. Rice, Steinmetz did most of his work on the calculation of alternating current phenomena after going with the General Electric company. His mastery of this problem did much to enable the General Electric Company to forge ahead until it commanded a position of leadership in the world of electrical affairs. As Mr. Rice has expressed it, "Steinmetz brought order out of chaos" in the matter of alternating current calculations.

"He abolished the mystery and obscurity surrounding alternating current apparatus, and soon taught our engineers how to design such machines with as much ease and certainty as those employing the old familiar direct current.

"It was fortunate indeed for our company and for the electrical industry that Steinmetz became associated with us at the critical time when the alternating current development had just started. It is not too much to say that his genius and creative ability was largely responsible for the rapid progress made in the commercial introduction of alternating current apparatus."

The great torch of Steinmetz's extraordinary talents was now blazing brilliantly, its far-spreading light increasing as time went on, and leading electrical men toward the approaching dawn of the world's electrical era. More indirectly, this great torch was lighting the way to a richer life for millions and millions of people, in America and throughout the civilized world.

CHAPTER XIII

STEINMETZ A CORPORATION EMPLOYEE

NOTWITHSTANDING the genuine eagerness with which Steinmetz, the young engineer possessing the masterly intellect, had accepted the proposal of E. W. Rice, Jr., to join the newly organized General Electric Company, he found it difficult to bid farewell to Yonkers. He never liked these breaks in life, as he used to call such occasions. And this was one of them.

"It was a break when I left Germany for Switzerland," he once told a group of friends. "It was another break when I came to America; and it was again a break when I left Yonkers and went to Lynn to join the General Electric organization."

His voice had a tinge of sadness in it as he spoke; and it was apparent that his ties at Eickemeyer's were not lightly broken.

There was another much deeper consideration at the moment. Not very long before, he had experienced one of the more heart-tearing breaks that come in life, the remembrance of which had not yet left him. This was the death of his father, which had occurred at the old home in Breslau about a year and a half previously. He had received word of it through a cabled message from one of his half-sisters.

Since the May morning when he said good-by to his father in the Breslau home, as he hurriedly prepared to flee from impending arrest, he had never again seen his parent; and only once or twice had he written home after coming to America. But the sorrow of this loss, when it came, was none the less keen.

It would seem that Steinmetz, in his inmost mind, had rather felt that Eickemeyer was, in a certain sense, much like a father to him during those happy years in Yonkers. Their relations, at all events, were very intimate, with a certain suggestion of filial regard on the part of young Steinmetz, particularly after that poignant cablegram arrived.

Beyond all this, his natural disposition was such that Steinmetz was always pleased to make friends; he was a great man for having cronies about him. The facility for keeping these friends was little less than a gift. Hence, he found it painful to say good-by to Eickemeyer and to the nondescript yet talented circle of Yonkers celebrities who had gathered with him those Sunday evenings at Seven Oaks.

Moreover, this was no mere departure to another locality. It was the complete transfer of the very company with which he had been associated. Eickemeyer was no longer in business at all but left after a while for a southern journey after a warm parting from Steinmetz. The latter stayed in Yonkers for some few months to superintend the transfer of the

Eickemeyer patents and engineering data from Yonkers to Lynn, which was for a time the headquarters of the General Electric Company. Then he also went to Lynn, finding himself assigned to the staff of H. F. Parshall, head of the General Electric Company's calculating department.

It is on record that Steinmetz went with the General Electric Company as a rank-and-file employee, valued above the average, to be sure, yet none the less an employee as distinguished from an official. He received a modest salary, compared to what he was eventually to earn, although it was good pay for an engineer in that period.

It is pertinent to observe at this point that he did not seek participation in the so-called official management of the company. In an organization where even the officials regard themselves as employees, Steinmetz seemed to take a quiet pride in remaining in the employee class beyond any question of doubt— one of the useful servants of the corporation, a valuable part of a great machine. This was his feeling, at least to a considerable extent, notwithstanding the premium that came to be placed upon his services, and the wisdom of the General Electric Company, both in giving him ample freedom of activity and in making it to his financial interest always to stay with it.

It is no less true that Steinmetz did not seek big pay. He might have tried his hand at bargaining for a fat salary when he perceived how eager Mr. Rice was to engage him. Instead of that, he agreed to the latter's proposal "without a moment's hesitation." It was the first instance where he really displayed his attitude toward personal wealth, the attitude of a pure-grained altruist, seeking only the best possible chance to do the work which fascinated him and which he believed would be of value to society at large.

And now began the golden age of electricity, which is even to-day hardly more than at the dawn. The thirty-year period between 1893, when Steinmetz joined the General Electric Company, and 1923, when he passed to the great beyond, saw an astounding forward stride in the electrical profession. But the biggest men in electrical affairs, then as now, recognized that much of this mighty development was due to the gentle-eyed, unostentatious, path-lighting torch-bearer of modern electrical engineering.

This remarkable attribute of Steinmetz was recognized perhaps more quickly by E. W. Rice, Jr., the General Electric engineer who secured the services of Steinmetz for the organization, than by almost any one else of that day. To Mr. Rice is to be credited the honor of linking up Steinmetz with a great company, which was itself under the helmsmanship of remarkable business leaders, distinguished for their sagacity and vision. The commercial super-genius was Charles A. Coffin, the young Lynn shoe man who deliberately entered the electrical manufacturing business because he foresaw that it was going to be the big business of the future.

CHARLES PROTEUS STEINMETZ

The General Electric Company thus began its amazing career of development well supplied with high-caliber human material for its commercial advancement. From the moment of its inception, its broad program was to sell the electrical idea to whole communities, entire cities, wide regions of the country; in short, the nation itself.

It likewise possessed exceptional engineering talent. Professor Elihu Thomson, of the old Thomson-Houston Company, was then, and has been since, an engineer of the first rank, a pioneer in research work and invention; and E. W. Rice, Jr., was destined to contribute much valuable work to engineering developments for which the company came to be noted.

The addition of Steinmetz was the acquirement of the keystone, completing an able engineering staff. Without the masterly guidance of Steinmetz, however, the General Electric Company's engineers would have been deprived of the torch they needed, and the company itself could hardly have taken the rapid strides forward which soon began to put it in a position of advantage. On the other hand, without the facilities which the General Electric Company gave him from year to year in the way of quarters and apparatus, the unhindered opportunity to do whatever work appeared at the moment of the greatest benefit to the general cause, and the constantly growing engineering personnel to whom to impart his mathematical methods and through whom to become more than ever noted in the profession, Steinmetz would have been a long time in stepping into his proper niche as a world-famous electrical genius.

Unquestionably, therefore, Steinmetz and the General Electric Company were of immeasurable help to each other—one of those instances where, in the course of time, each made the other famous.

The calculating department of the General Electric Company at Lynn, where Steinmetz now took up his work, handled all the mathematical computations involved in the design of the electrical apparatus that the company was manufacturing. Hence, it was the one department where Steinmetz belonged and where, as it happened, he felt at home from the beginning.

That particular period constituted the infancy of electrical practice. The potential power of mighty Niagara was just beginning to be put into service. The General Electric Company had only just completed its River Works at Lynn, which was not as extensive as it is at present.

In the calculating department, under Mr. Parshall, was done all the pioneer engineering and all the designing for every class of apparatus, no matter for which field of service intended—work which to-day is highly specialized and scattered through several departments, classified by the character of the machines.

Steinmetz quickly found new cronies among the

young engineers who worked under Parshall. The staff numbered all told about nine men. Those who became the most congenial associates of the brilliant new-comer included E. J. Berg, who, with his brother, Eskil Berg, was to share with Steinmetz, a year or two later, in one of the most unique semi-Bohemian domestic establishments that have ever given shelter to a group of scientists.

The first engineering problem which Steinmetz was asked to pass upon after he became a General Electric engineer was in relation to electrical machines for the first power-plant at Niagara Falls. The General Electric Company was preparing a proposition for this power-station, which at the moment was the biggest job engaging the attention of electrical men. There was nothing especially remarkable about the proposed apparatus—at least not to Steinmetz—although his associates were immediately impressed by the readiness with which he handled the mathematical calculations involved and the lightning-like swiftness with which he obtained his results.

Steinmetz next made an investigation of the transformers built by the company, to see how they could be improved. He was unquestionably exercising his keen facility for looking ahead when he took up this line of work; for he foresaw with the utmost clearness just how electrical expansion was to proceed, in a general way, and how indispensable to that expansion the transformer was to be.

During this period, also, he studied the induction-motor, which was then just being designed. Working with Dr. Louis Bell, he conducted tests that led to the redesigning of these motors and the production of an improved type.

It was now about the middle of 1893, or perhaps around September; and the whole work of the department, indeed, the work of the entire plant, began to be seriously cut down by the disastrous business panic of that year. After a few weeks the times became so dull that there was virtually no work for the calculating department to do.

The money-market was naturally affected, so that, although the General Electric Company was doing a very good business considering its youth and the stringency of the times, it experienced considerable difficulty in doing its financing. Steinmetz and the rest of them were not always positive that they would receive their money on pay-day; but the company always succeeded in meeting its obligations to its employees, although it had to pay a high rate to obtain money.

Morning after morning, Steinmetz reported for work but was told that no work had come in. No one was buying electrical apparatus; hence the company was not designing or building any new machines. So Steinmetz sat around in Parshall's office, and he and Parshall, with W. C. Fish, another of the engineers, spent much of the interval smoking, telling stories and cracking jokes. Frequently they put in hours at a time in this manner.

Meanwhile, the shops were idle at least half the time. About a hundred men were working three days every second week. Parshall's staff was cut down until he had only three men left, and these three worked but three days a week. They were E. J. Berg, E. B. Raymond, and Steinmetz.

The General Electric Company, although only a year old at the time, had a creditable exhibit at the World's Fair in Chicago in 1893. Perhaps the most prominent part of the exhibit was the Intramural Railway, a third-rail system which ran on an elevated structure. Motor-cars hauling trailers constituted the trains. This was probably the first elevated electric railway in America.

A party of engineers was sent to Chicago by the company to superintend the exhibit. The group included Steinmetz, who was alert to see everything that was in any respect a new development.

He experienced while there a pleasant surprise upon running across an old friend of his Yonkers days, none other than Tischendorfer, "the Dutchest Dutchman." When Eickemeyer sold out, Tischendorfer had gone back to Germany, where he became prominent in the Schukert Company, now the Siemens-Schukert Company, manufacturers of searchlights. At the World's Fair this concern had an exhibit of searchlights, of which Tischendorfer was in charge.

Steinmetz was greatly amused by observing the good time Tischendorfer was having playing various small pranks; in his reception-room beneath the floor of the main exhibition-hall he entertained with sausages and German wines and obtained illumination by tapping the wire that supplied current for a neighboring exhibit, much to the perplexity of the owner of the exhibit.

The welcome renewal of acquaintance with the erratic Tischendorfer and the equally welcome additions to his list of congenial friends gave to this year 1893 much that appealed to the peculiar social disposition of Steinmetz. In rather pronounced contrast, however, was his considerably less congenial leisure life during his residence in Lynn.

His living-quarters consisted only of a room in a boarding-house at the corner of Commercial and Common Streets. There was a grocery store on the ground floor, and many other small stores and rooming houses in the neighborhood.

The husband of the landlady kept a small second-hand furniture store. His great weakness was an over-fondness for liquor, which led him to stay out until the small hours of the morning and to make so much noise when he did come home as to awaken nearly everybody in the house. Steinmetz did not enjoy this, although he found the landlady herself most agreeable. His departure for Schenectady came before he had been impelled to find another boarding-place where his slumbers would not be periodically disturbed.

After Steinmetz went to Schenectady, however, the

times became so very bad in Lynn that his erstwhile landlady decided she, also, could do better in the new headquarters city of the General Electric Company. She left her husband in Lynn, which proved a good thing for her new enterprise. But after an interval, the husband showed up in Schenectady, whereupon business began to fall off at the boarding-house. Steinmetz had a room there for a while before he became permanently settled in the Mohawk Valley city, although he lodged for the first two weeks of his residence in Schenectady at the Edison Hotel.

In December, 1893, word was passed around among the men of Parshall's department that the entire department was to be transferred to Schenectady. The news was a bit dismal to Steinmetz, for it forced upon him another one of those unpleasant breaks.

With all his sagacity for predicting the trend of electrical development, he could not apply the faculty of foresight to his own affairs, or else he would have discovered that this was the last break he would be obliged to undergo.

The transfer to Schenectady afforded him an interval in which he could return to Yonkers for a Christmas visit with his old friend and former fellow-worker, Edward Mueller, draftsman at Eickemeyer's factory when Steinmetz first went there, and with whom he had lived during most of his residence in Yonkers. This was not his only return visit to Yon-

kers, for several times after he went with the General
Electric Company his former employer, Eickemeyer,
sent for him to seek his advice on some question or
invention with which the latter was working. This
continued until the death of Eickemeyer, in January, 1895, at Washington, while on a trip to the
South to recover his health, which was poor during
the last few years of his life.

Not all the men who were with Parshall in Lynn
were included in the group that was transferred to
Schenectady. The transfer was effected in January,
1894, and it brought Steinmetz into considerably
greater prominence than he had previously experienced. The principal reason for this was the departure of Parshall for England, where he was sent
by the company to conduct extensive tests of some
large alternators which were not operating satisfactorily. Subsequently he became an independent
consulting engineer in London.

At Schenectady, Steinmetz soon was conceded to
be the one great indispensable person among the engineers who were engaged in alternating current
work. The recognition which was his from the moment he joined the General Electric organization at
Lynn was increased many-fold as soon as he established himself in his new environment. By tacit approval and official decree, he became the head of the
reorganized electrical calculating department, which
particularly concerned itself with the calculation and
design of alternating current machines.

CHARLES PROTEUS STEINMETZ

Every engineer who worked under Steinmetz in the next four or five years depended upon the "little giant" to show him the symbolic method in making calculations with the alternating current, so that the calculating and designing department in reality resolved itself into a sort of glorified school-room. Steinmetz was the schoolmaster; and never was there a more discerning, painstaking, or patient school-master. Many folk who knew Steinmetz at various times in his life have agreed that they never once saw him exhibit the least degree of impatience, nor the slightest discourtesy toward those who sought his help in the mathematical work of the office.

By the class-room method this mathematical genius of electrical engineering gradually demonstrated the working of his symbolic method until he had made all the General Electric engineers familiar with it. There are men still on the engineering staff of the General Electric Company who gained their first knowledge of this procedure from Steinmetz in the days of the old calculating department.

"We became pretty familiar with the symbolic method at length," said one of Steinmetz's "boys"; "but we never really mastered it. We could never do the things with it that Steinmetz could do. Steinmetz grasped a thing all at once, even an intricate mathematical problem.

"E. J. Berg, who worked as his assistant, used to say that Steinmetz could take any equation up to the third degree and visualize it out in space. There is not one man in a thousand who can do that.

"Yet he was not arrogant in the smallest degree. He never 'showed off' his knowledge, or tried to glorify himself. He was one of the most modest of men, and realized more than any one else that he did not know it all. He showed this in the way he would answer an inquiry. He never considered himself infallible, and hence would never give an answer that suggested finality. He would say, 'That looks all right; we 'll try it.' He would tell you whether a proposition appeared sound or not, but he would not assert, with conceited positiveness, that it absolutely was, or was not, correct, as though he were a divinely appointed oracle in whom there could be no error."

Among the well-remembered feats in mental mathematics performed by Steinmetz and preserved from this period was the question that somebody raised in the course of the day's work as to the cubic content of metal which is removed from a cylindrical rod two inches in diameter when a two-inch hole is bored through the rod, separating it into two pieces. By all ordinary methods, obtaining the answer to this proposition would require considerable time. But when it was laid before Steinmetz he merely considered for a moment, and in less than two minutes he had worked it out in his head, quietly stating the answer.

From 1894 to 1898, Steinmetz and his corps of assistants had the entire burden of alternating current calculation and design. During most of these years the department was housed in what is now Building 4 of the Schenectady works, later going to the front of Building 6, where it was located several years longer, and where Steinmetz remained until he moved into the handsome new general office structure, designated as Building 2.

The ultimate organization was as follows: Charles P. Steinmetz, engineer in charge; E. J. Berg, assistant on general work; A. E. Averett and Walter I. Schlichter, assistants on induction motor work; Eskil Berg, assistant on synchronous motor work.

Every indication points to a profound realization on the part of all these men, and all others who had occasion to work with Steinmetz, that they were associated with a veritable genius in their own particular field. It is apparent that the engineers of the company were amazed day after day by the almost uncanny facility of Steinmetz for immediately solving the most intricate mathematical problems without recourse to paper and pencil. He had made a deep and lasting mark upon the profession of which he was a member—evidenced, if in no other way, by the utter inability of the rest of the staff to work out their alternating current calculations without falling back upon the author of the symbolic method.

And in this year 1894, the year of his removal to Schenectady, Steinmetz took the step that consummated his naturalization as an American citizen. The necessary five-year period since his entrance into the country in 1889 had now elapsed, and hence he was eligible, under the law, to apply for his final papers.

Steinmetz, therefore, journeyed back to Yonkers on the proper day, appeared before the court, and was made a full citizen of the United States, his willingly adopted country, acquiring citizenship in the speediest possible time allotted for such proceedings. And those who knew him best discerned much quiet satisfaction in his mind over his acquirement of full American citizenship.

He cast his first vote as a naturalized citizen at the municipal election in Schenectady in the autumn of 1894.

The building at Schenectady in which the electrical calculating department was first quartered made no claims to being fire-proof. In Building 4, the accommodations were merely temporary, while alterations were being made in the building at large. Rough wooden partitions were all that divided the various offices from each other.

Because of this, and because the company considered that a general policy of conservation was wise, as the number of its employees began to increase and the safe-guarding of its property from accident assumed greater importance, a "no-smoking" policy was inaugurated. The announcement of this decree was conveyed to the engineering force by means of prominent placards stating that smoking was not permitted in the offices. These posters made their appearance very soon after Steinmetz had become established and his department had begun to function.

Now, Steinmetz, by this stage in his career, had become a firm believer in smoking. Already he and his cigar were absolutely inseparable; not a day passed, and not a moment in any day, but that Steinmetz was to be seen with a long, thin cheroot in his mouth. No one could remember having observed him without it, nor could Steinmetz himself conceive of working without a haze of blue smoke about his head.

Hence the dramatic possibilities that were apparent when Steinmetz first caught sight of the neatly lettered cards which placed the official ban upon the use of tobacco during working-hours!

As it happened, however, no temperamental incident developed, either then or at any later time. Neither was there any cessation of smoking by Steinmetz. So far as any of his associates of the period can remember, the "little giant" merely made some half-whimsical comment, the burden of which was that he could not work without smoking and that therefore he intended to continue smoking.

Thereafter he apparently gave no thought to the signs but went on with the investigations and calculations of the moment. No one took him to task for smoking in defiance of the official dictum, for it was daily apparent to every official of the company that without Steinmetz the General Electric engineers would be virtually helpless in trying to handle the alternating current by the symbolic method.

And as time went on, Steinmetz showed himself more and more valuable to his employers; hence he was permitted to smoke in his own office and laboratory with an ever-decreasing probability that he would be told it was against the company's rules.

As the circumstances of those days are recalled, it appears that Steinmetz was not alone in his course of action. Most of the other engineers of the company also chose to ignore the no-smoking rule, although few of them intimated that they might have to quit work if unable to smoke.

This incident was the origin of perhaps the most famous Steinmetz anecdote in existence. The story was related, in the innocence of its original occurrence, by some of the engineers to their friends. Then it began to spread among other departments in the General Electric works. The propensity for exaggeration soon got in its work; and as time went on, the episode became so universally known in Schenectady at large that to this day it is a household legend throughout the length and breadth of the Mohawk Valley. Few remembered, or ever knew, after a while, that Steinmetz was not the only engineer who declined to be bound by the anti-smoking decree. Nor did his quiet, unspectacular protest figure in the occurrence as it came to be popularly told and retold, neighbors relating it to each other, residents recalling it for the benefit of out-of-town visitors, storekeepers recounting it to their customers, and the customers to their friends. Every smoking man seemed to seize upon it as a classic and to feel a hidden delight over the discovery that at last there had come forth a worthy champion of the whole great fraternity of smokers.

In the endless succession of repetitions, through a long stretch of years, the incident finally became so badly twisted around that Steinmetz was represented as a tobacco devotee who stayed at home rather than work without his cherished cigar; and the exaggerated version of the affair described with impressive detail how they missed the "little wizard" when a complicated mathematical problem came up, which nobody else could do anything with; and how a "high official" of the company hurried to Steinmetz's home to find out what the matter was, receiving the gentle ultimatum: "No smoking, no Steinmetz." The earnest solicitation of the official that Steinmetz return and the eager assurance that he would be allowed to smoke without let or hindrance was portrayed as the closing scene of this narration of the happening.

As this memorable little tale became better and better known, the good folk of Schenectady came to venerate Steinmetz's cigar very nearly as much as the man himself; and on the day of his death this was the first thing that some of them, at least, remembered about him when they heard the news.

But they recalled it more as an expression of

kindly esteem than with the slightest thought of disparaging his real life achievements; and it was all a tribute to the deep human appeal of the man, the evidence that, after everything was said and done, he was as human as any of us, with his likes and dislikes, his hopes and his anxieties. And so, in every possible sense, he really was.

CHAPTER XIV

THE LIFE OF A BOHEMIAN SCIENTIST

POSSESSING curious social instincts, which led him to seek out liberal-thinking, self-reliant chums among his daily associates, Steinmetz displayed his fondness for intimate companionships more and more as he advanced in the estimation of his fellow-men. People who did him a service, however small, those who merely showed thoughtful human interest in him and his affairs, were apt to find themselves rewarded with his permanent friendship. If these people were thrown in with him by the common bond of professional interests or by daily work together, they were likely to become his lifetime intimates, his room-mates or housemates, in short the counterpart, with Steinmetz, of a domestic circle.

It was thus that one of the most striking, perhaps the most winsome, of all his human relationships blossomed into full flower. As every one knows, the last period of his life was remarkable for a wonderfully happy family group, in which Steinmetz was the center of the picture, first as "dad" and then as "grand-daddy," by adoption, to a whole array of young folk.

A BOHEMIAN SCIENTIST

And yet the beginning of this peculiarly delightful association was in some comparatively insignificant occasion, some forgotten interest which was displayed in his welfare, or some chummy disposition casually uncovered, which literally won the heart of Steinmetz and led him eventually to adopt, as his own son, Joseph LeRoy Hayden, who was closer to Steinmetz than any one else in his later life. Hayden himself was led to wonder at it all, having no definite consciousness of any single reason for the development of this intimacy. The beginning of it was so hazy that it would be impossible to ascribe it to any tangible incident.

Hayden had not yet come into Steinmetz's life when the latter arrived in Schenectady, in January, 1894, his feelings rather depressed over this new and swift change in surroundings. He viewed with no great anticipations the old barge-canal, the winding Mohawk, and the low, flat tract on which the buildings of the General Electric works were huddled in an unpretentious group. His one consolation in this whole dispiriting business of changing his headquarters was that some of his Lynn cronies had come along with him.

Of these companions, E. J. Berg was at that time the first and foremost. The two of them were very much together outside of office-hours. They thoroughly explored the town, walking frequently up and down State Street, where Steinmetz, a queer, gnome-like figure stepping quickly along in the company of his stalwart comrade, attracted constant attention.

After a week or so, the two chums, having got into the new routine and finding time to carry their observations further, discovered that Schenectady had a few compensations about it, after all, the principal one of which was the Mohawk River. When they became aware of the possibilities offered by this rambling, placid stream, with its low banks and distant hills, they decided that they were going to like their new home.

Forthwith the pleasant occupation of exploring the river absorbed their leisure for many weeks. With the approach of spring, their activity increased. They were not content merely to walk along the river banks; they agreed that it was absolutely essential that they should have a boat. Furthermore, it would have to be a boat built to order.

After considerable difficulty, they persuaded an old river-man named Joiner, who had a great local reputation, to undertake this task. Joiner, although a skilled craftsman, had long since retired from business; hence he was reluctant again to take up his tools; he was even stubborn in his refusal. Eventually, however, the engineers persuaded him that they were serious and that they would pay him well, Steinmetz acting as spokesman in the deal. The boat was thereupon built, proving, when it was finished, all that they could desire. This boat finally became the sole property of Steinmetz, as he bought out the interest of E. J. Berg.

A BOHEMIAN SCIENTIST

The acquisition of the boat intensified their interest in the river and its scenery. Long trips up or down the stream came to be the favorite diversion on weekends and holidays. And one of the first of these, on a Sunday afternoon early in the summer of 1894, had the effect of converting Steinmetz into a permanent lover of the Mohawk.

This excursion carried the party some miles above Schenectady, as far as the mouth of a small tributary stream locally known as Viele's Creek. Attracted by the pleasant vista at this point, they rowed into the mouth of the creek, proceeding further and further, until the boat suddenly ran aground on a gravel bar.

Two of the party clambered out and waded ashore, wandering along the bank on foot. Steinmetz and Berg stayed in the boat, which had floated off the bar after being lightened. By intermittently rowing and pushing, they boated a considerable distance, finally landing and climbing to a delightful spot on a high bluff, from which they could see the quiet river, with here and there an island, and beyond the hazy outlines of the hills.

There the land party found them somewhat later, admiring the outlook and eating their lunch. All agreed that the spot would make an ideal site for a camp. Little else was said upon the subject, and no one imagined that there would ever be a camp actually established on the spot.

Five years later, however, there really was a camp

on that site; and it was the camp of Charles P. Steinmetz, the ardent admirer of the Mohawk.

Long before Steinmetz had completed his first year in Schenectady he felt thoroughly at home there. He had ceased living in hotels and boarding-places; instead, he and his partner, E. J. Berg, had rented a house. It was the beginning of a jolly, fraternal, free-and-easy mode of life, with many Bohemian touches about it, a bachelors' hall with ultimately a trio of bachelors as inmates, each of whom was free to indulge the idiosyncrasies of his bachelor nature.

It might appear, from a chronicle of that curious domicile, that the idiosyncrasies were rather freely scattered among the inhabitants; yet, for sheer whimsicality, Steinmetz probably outdid them all. It was he who paid the least attention to conventions of dress; it was he who was most apt to undertake some chemical experiment in the house, with the probability that woodwork or furniture would become blotched with stains, or, if it was a particularly daring experiment, the retorts might blow up; and it was he who managed to gather unto himself the nearest approach to a private zoo that the staid descendants of Schenectady's Dutch settlers had ever beheld in their midst.

It would have astonished the business and professional men whom Steinmetz encountered in his engineering work, the business promoters of the General Electric Company, and the public that had just begun to hear, in a hazy manner, about young Steinmetz, could any of them have seen what went on in the private dwelling-place of this master of things mathematical "after hours." The man who was the brains of many electrical achievements was a leading spirit in a group of fellows who complacently shifted for themselves, yet maintained a fairly pretentious domestic organization, strictly on a partnership basis, with a woman housekeeper and cook, and sometimes an extra maid of all work.

As like as not, at any moment, they would slight all domestic operations if some fascinating electrical subject was absorbing their attention. This was particularly true of Steinmetz.

Naturally, it was not long before some of the peculiar habits of the odd-appearing young engineer began to be noised around the town. People told of seeing him repeatedly walking the streets without a hat, and sometimes in cold weather without an overcoat. It was recounted that he thought nothing of wearing an old sweater, or ill-matched trousers and coat, when in the presence of prominent officials of the General Electric Company.

These incidents, in time, became the originals of many of the best-known Steinmetz anecdotes, some of them as proverbial as the smoking story. The incidents themselves were not actually numerous; the same cannot be said of the anecdotes that have grown out of them.

The first of the chummy domestic establishments

in which Steinmetz was a figure was located in a rather small, plain-looking house on Washington Avenue, a comparatively short walk from the General Electric works. Here Steinmetz and Berg, like two kindly, sociable souls, conducted a masculine community of their own that was not unlike the rooming life of Steinmetz and Asmussen several years before in Harlem.

About 1897 the house on Washington Avenue was relinquished for a larger one at the corner of Liberty and College Streets, where the partners set up a more elaborate residence. A year later the bachelor circle was enlarged by the arrival from Europe of Mr. Berg's brother, Eskil Berg, who came to Schenectady to join the General Electric staff of engineers, and who thenceforward became a harmonious unit at the Steinmetz-Berg fraternal fireside.

So unrestrained was the life of this household, as regards self-imposed rules or communal restrictions, so completely was each one left to follow his own whims and fancies, always retaining a healthy consideration for the others, that the amused friends of the trio came to allude to the home of these original souls, with recognition of the appropriate geographical location involved, as Liberty Hall.

Certain it is that Steinmetz, for one, felt free to indulge several characteristic hobbies. The first of these was his fondness for rare plants, especially certain varieties of cacti. He had a small conservatory built to adjoin the house, and there he installed several specimens of this plant, which he had purchased with the thought of starting a collection. He soon had a respectable showing of growing things, which he cared for with much patience, and which he delighted to show his friends.

Later additions to the collection consisted of some very fine orchids, always a favorite flower of his, and a number of different kinds of ferns.

Soon, also, he and the Bergs began to acquire all sorts of curious pets. Most of them were bought at first by E. J. Berg, who cared for them and saw that they were fed. Some, however, were the special property and particular delight of Steinmetz. They were all kept in the back yard, where various pens and cages were constructed for their accommodation.

Of the whole array of pets, Steinmetz had an exceptional attachment for a couple of jet-black crows, with which he had made friends in some unaccountable manner. And fast friends they were, Steinmetz and the two crows, so much so that they would fly down into the yard to be fed, when they saw him come out of the house, and would perch on the sill of his bedroom window when they knew he was in the room.

Steinmetz named them John and Mary and went through the formality of conversation with them just as if they were human beings. In fact, he coincided with Ernest Thompson Seton by averring that one can understand the language of crows if one will only take time and pains to learn it. He always maintained that he could tell perfectly when John and Mary wanted something to eat, or when they were provoked over anything.

It is unquestionably true that these crows really did identify Steinmetz to the extent of flying into the yard to "chat" with him, and locating his bedroom by peering in all the windows until they discovered him.

The friendship between Steinmetz and the crows continued, to the great amusement of the neighbors, throughout the lives of the birds. This was not as long as it might have been, because of an unfortunate occurrence. For one of the crows met a tragic end; which one, the historical records of Liberty Hall fail to disclose. It was while the two crows were flitting about the menagerie yard, awaiting the appearance of Steinmetz in fancied security, that a pet raccoon confined in one of the pens got loose in some manner, surprised one of the crows, and slew the hapless bird on the spot. He had even begun to eat his victim when the crime was discovered and the culprit driven off. The other crow did not survive this sorrowful event, while Steinmetz was much grieved over the deaths of both of them. He had them stuffed and kept them the rest of his life, assigning them a prominent perch on a high bookcase in his Wendell Avenue home.

The taxidermist, to Steinmetz's unending satisfaction, preserved a decidedly chipper look about the crows, so that Poe's raven would hardly have felt at

One of Steinmetz's queer pets. John the Crow, looking in the window of Steinmetz's room at Liberty Hall

Steinmetz's favorite dog, Buck, wearing a cap and smoking a pipe. Photographed by Steinmetz at Wendell Avenue, on the lawn

ease in their presence nor have found much provocation for his mournful refrain of "Nevermore!"

The list of birds and animals which Steinmetz and the Bergs finally collected at Liberty Hall at length became so impressive that the fame of the place spread throughout the neighborhood. Mr. Berg added several cranes and a few young eagles. There were also some owls, which were much thought of by the men, although at times they frightened the servants by hooting in the night.

Among the special favorites was an unusually intelligent monkey named Jenny, who was a source of continual amusement. She stayed with the men until their circle began to break up, when she was given away. Squirrels and dogs were numerous at different times.

There were several raccoons, more or less tame. One of these was the chum of all the men and seemed such a knowing little chap that he was allowed to roam all over the house until the cook objected that he was helping himself to things from the larder. After that he had to stay inside his pen.

Finally there was a lusty three-foot alligator, in which Steinmetz took a special interest, finding delight in showing his friends how he could make the alligator do almost anything he wanted it to do. By this time the renown of the menagerie was widespread, and the men frequently invited schoolchildren in to see the sights.

CHARLES PROTEUS STEINMETZ

Steinmetz was very fond, at this period, of performing laboratory research experiments at all hours. He always liked to have a home laboratory, and there he was to be found working long after he had left his office at the plant. Most people would have thought Steinmetz overworked if they had seen him spending whole evenings in this manner; but it was not work to him. Electrical engineering to Steinmetz was the one fascinating delight—something he wanted to be busy at all the time.

His Liberty Street laboratory was begun in his own room but soon expanded to such an extent that it was moved out to the stable. Some of the simpler experiments, however, were still performed in the house for the sake of convenience; and it must be observed that most of the time Steinmetz was eminently successful in his uncertain venture of mixing home life and laboratory activity.

There was no lack of tranquillity, as a general thing, in Liberty Hall. Still, it must be confessed that once or twice the chemical operations of Steinmetz were of such an unexpected nature and produced such harrowing consequences that they led the housekeeper to enter astonished protests. This happened when a retort was upset, spilling the liquid contents upon the carpet, where they left a large stain; or when chemicals were distilled which left an almost unbearable stench about the house. Feminine patience was especially tried one evening when the flame from a laboratory burner set fire to the room, giving the whole house a few minutes of very real and unwelcome excitement.

During the winter of 1895–96 one of Steinmetz's half-sisters, Miss Clara Steinmetz, came for the first of several visits. She assisted in some of the housekeeping duties during this stay, and more so during a later visit when Steinmetz was the only one of the trio remaining in the house.

She was made welcome, although from all accounts the mixture of home life, laboratory activity, and scientific discussions, with its pronounced atmosphere of masculine peculiarities, did not greatly appeal to her. She did not see the logic, for instance, of harboring such an array of meat-eating animals as to result in a monthly bill for food for the pets that was sometimes larger than the bill for feeding the human beings.

Moreover, Miss Steinmetz was taken sick during the winter, and the young engineers of Liberty Hall, in boyish roguishness, pretended among themselves that she was suffering from scarlet fever. Whether or not they intimated this to her is not clear, but at least they had the house for a few days in a state of mild consternation; and let it be understood that the whole establishment was in quarantine.

Miss Steinmetz was an artist of some talent (a little later she painted a portrait of her brother, which long hung in the Wendell Avenue house), and being attracted to New York as a favorable atmosphere for artists, she left Schenectady in the spring, to live in apartments in the metropolis.

The happy-go-lucky life at Liberty Hall continued at an accelerated pace. About this period, as nearly as can be determined, Steinmetz and the Berg brothers organized their weird Mohawk River Aërial Navigation, Transportation, and Exploration Company, Unlimited. The "unlimited" aspect of this fun-promoting organization came from the circumstance that any number of members could gain admittance, on the sole qualification of possessing a two-dollar entrance fee. There is no authentic record as to the size of the membership, but it is indicated that "a large number" of General Electric engineers was enrolled.

The particular object of the Mohawk River Aërial Navigation, Transportation, and Exploration Company, Unlimited, was the construction of air-gliders. In this worthy enterprise they displayed a magnificent anticipation, by a few years, of the gliders that have been developed to-day as a successful type of aërial transportation in Europe. Three or four gliders were actually built under the supervision of Steinmetz and the Bergs.

The machines were given practical tests on the hills at Hoffmans, a small suburb of Schenectady on the Mohawk River. But they never would stay up in the air. As E. J. Berg explained it, "we did not get just the proper construction of the wings." The company was a serious affair, however, to the extent of actually planning to market and exploit the gliders, if they could ever be made to glide.

Steinmetz was the embodiment of enthusiasm over this whimsical side-project. He came out to the tests with a camera, prepared to make a photographic record of the experiments. Not to be denied this culminating satisfaction, he did take several pictures of the gliders while the machines were resting on the ground. Then he retouched the negatives to make it appear that the gliders were in the air; and the doctored photographs were solemnly exhibited to a number of the wondering members of the company, who for a while were allowed to cherish wild dreams of a tremendous business development when the inventors began to publish their great discovery.

Several whimsical incidents during this and the succeeding years at Liberty Hall reveal the Steinmetz character in all its spontaneous boyishness. His passionate devotion to cigar-smoking would almost serve to discount the remarkable contest in abstention in which he and E. J. Berg engaged, were it not vouched for by Mr. Berg himself. The two had fallen into a discussion at dinner one evening as to the possibility of exercising sufficient self-restraint to abandon any habit of long standing. Steinmetz thought it was quite possible, but Berg was not convinced. Finally, Berg averred that there was no way of finding out except by a practical test, in true engineering style; and he accordingly proposed that

Steinmetz and himself should both refrain from smoking for a period of a year.

Steinmetz immediately agreed. They called Berg's brother, Eskil Berg, to witness, and both proceeded from that day forward to leave all forms of tobacco strictly alone. Strange as it may seem, they kept to the agreement without a break. For twelve months the engineers at the General Electric works witnessed the extraordinary spectacle of Steinmetz working minus his cigar. Some of them must have thought the mathematical master had undergone a change of heart about observing the no-smoking decree.

When the date arrived that marked the termination of the year, Steinmetz was in Cleveland, on a business trip, while Berg was in Schenectady. Early in the day, Berg despatched a telegram to Steinmetz to this effect "Year is up, and I'm smoking for all I'm worth." He had scarcely got the telegram off when a telegram from Steinmetz was handed to him, sent about an hour previously and conveying the same message almost to a word.

It is recalled by Steinmetz's associates that he never had the taste for beer which was common to most Germans. He would sometimes take a little with his meals, and now and then would drink some wine, but he was rather noticeably un-German in the matter of his beverages.

He was quite fond, however, of a very fine Jamaica rum which was sent to him by his friend, William

Stanley, of Pittsfield. In fact, he aroused E. J. Berg's displeasure by appearing to take advantage of Mr. Stanley's generosity when the rum was gone to the extent of hinting that he would be glad to have another supply. Mr. Stanley promptly sent him several gallons more.

Berg decided that this was going too far; so he took Steinmetz to task about it, contending that he could get just as good rum by buying it and that it would be better to replenish their store in that manner. Steinmetz held that there was no better rum to be found, whereas Berg declared that Steinmetz would not know the difference if he tasted a really inferior brand.

To prove his point, he slyly watched when Steinmetz drank rum from the decanter on the table, and for each glass of rum that Steinmetz poured out, Berg secretly put into the decanter a spoonful or two of strong beer. Steinmetz never perceived the difference in the taste of the rum, until at length Berg was able to prove to him that he was drinking about ninety per cent poor beer and ten per cent good rum, without having the slightest inkling of it. This little trick, it is related, greatly shocked Steinmetz, who considered it far from courteous. But he said no more to Mr. Stanley on the subject of rum.

Steinmetz had the fraternal esteem of his associates —the Bergs and the many engineering friends who dropped in to visit them almost nightly—to such an extent that he was elected president of a whimsical sort of club, which was long kept up, known as the Society for the Equalization of Salaries. It held weekly sessions every Saturday night.

The bond of friendship between them all was in reality of the very finest and strongest; and Steinmetz was happy in the good graces of several women of the neighborhood, in addition to his boon companions among the men. For, contrary to some unfortunate misapprehensions concerning him, Steinmetz was as much pleased to have the esteem of women as of men, and he was always a capital comrade to small girls.

The neighbors who most frequently dropped in to see him personally were the Kruegers, who lived a little further down on Liberty Street. Mrs. Krueger and her two children, Carl and Gretchen, were keen admirers of Steinmetz; and in their society he found much congenial interest.

Substantiating this impression of the genuine appreciation which Steinmetz felt of feminine charms, and directly contradicting the notion that he was over-bashful toward women, Berg records that Steinmetz had several most cordial friends among the fair sex, while, as for children, Berg remarks, "He remained a child in spirit all his life, and loved children and young people generally."

In the autumn of 1900—it was October 20, by Eskil Berg's diary—the gradual break-up at Liberty Hall began. Eskil Berg himself moved away on that date to another part of the city. E. J. Berg continued to live with Steinmetz for a time; then he, too, went elsewhere in Schenectady to live, and Steinmetz stayed there for a time alone.

He continued as the sole permanent occupant of the Liberty Street house, except for one or two visits from his sister, Miss Clara Steinmetz, until well into the spring of 1901. But he was by no means left in complete solitude. Various acquaintances among the young men at the General Electric plant called frequently in the evenings, and he also saw considerable of his friends, the Kruegers.

Early one evening in that winter, not long after the Bergs had moved away, two young men dropped in for a brief visit. One of them was an acquaintance of Steinmetz's named Chamberlain, whose social inclinations led him to stop in to see Steinmetz whenever opportunity offered.

The other was Chamberlain's room-mate, a young engineer at the General Electric works who, up to that evening, had never seen the brilliant mathematical worker. It was Joseph LeRoy Hayden, who eventually was to become Steinmetz's foster-son.

The two were shown into the house by Miss Steinmetz and conducted up-stairs to a large hallway, on the second floor, which served as a reception-room. From one of the rooms opening out of this hall they presently saw Steinmetz emerge and walk toward them. He greeted Chamberlain in his customary quick, eager, alert manner, and shook hands with

Hayden, whom Chamberlain introduced. Then they all sat down and visited for a while, talking about a variety of subjects, until the young men took their leave.

Hayden carried away with him an impression of a singular personality. He had observed with inward wonder the sight of this little gnome of a man, whose head seemed large out of all proportion to his body, whose face was kindly, with much hidden keenness about it, and who looked young despite the luxuriant dark gray beard. His manner seemed to Hayden to be reserved almost to the point of shyness.

Perhaps it was that very reserve that led men to seek a closer acquaintance with Steinmetz; for no one ever felt that he knew Steinmetz perfectly after only one meeting, nor indeed after many meetings.

At all events, Hayden was eager to meet Steinmetz again, so that Chamberlain had no difficulty in persuading him to make other calls. Steinmetz, on his part, ever responsive to those who took more than a passing interest in him, seemed to welcome this opportunity for an intimate friendship with a relish that suggested the yearning he always felt for congenial companions.

And so Hayden became one of the group of young fellows who knew Steinmetz well enough to drop in once in a while, assured that they would always find a welcome, and aware that Steinmetz was a man worth talking with. Hayden and Steinmetz liked each other almost from the first, although there was no definite acknowledgment of friendship; it was simply a slow-growing process, which for a long time exhibited no unusual characteristics. Hayden kept coming more and more frequently, however, until he was to be found at Liberty Hall more or less every evening.

Steinmetz, throughout this interval, lasting until the end of the winter of 1900–01, was taking his meals at Mrs. Krueger's hospitable home but living in the old Liberty Hall house, maintaining his small conservatory, and keeping a good many of the array of queer pets that had been collected while the Berg brothers lived with him.

Just at this time, however, Hayden heard him talking a good deal about a real-estate venture in which he was interested. He had been attracted by the opportunity to purchase land and build a house of his own through the formation, a short time previously, of the General Electric Realty Company. This concern had purchased, for restricted residential development, a large stretch of undeveloped property in northeast Schenectady, beyond Union College, for the purpose of reselling it to persons connected with the General Electric.

Steinmetz realized that he needed more of an establishment for his dwelling-place than he could ever secure in a rented house. This conviction had been borne in upon him rather sharply by a fire that burned down the stable in which he had his laboratory, consuming some of the laboratory equipment and leaving him almost wholly devoid of one of his chief delights. This unfortunate event had occurred before Hayden first called upon him; but it was still a topic of conversation with Steinmetz, who discussed it with his friends, and also told them about his plans for buying property in the General Electric realty tract and building his own house.

As spring drew near, Hayden heard much discussion during his evenings at Liberty Hall about the Steinmetz camp, which had only been constructed the previous year. Steinmetz had leased the site on which, five years before, he and his party of friends had stopped during a Sunday excursion to eat lunch and admire the view. He had engaged a farmer who lived near-by to do the actual building of the camp, which in the beginning was simply one fairly large room with a gabled roof, the front half supported by timbers, so that it looked like a little hut on stilts.

The camp had acquired something of a reputation the previous year among the young fellows who knew Steinmetz well, for invariably they were invited to "come out to camp" every week-end of the spring and summer. Hayden also received this inevitable invitation in the ordinary course of events, until almost every Saturday and Sunday found him at the rough-and-ready little lodge that nestled among the trees on the bluff of the creek; and there he was drawn closer than ever into the circle of Steinmetz's friends, to the growing satisfaction of both.

The camp on Viele's Creek was so very new that Steinmetz was not yet satisfied of its stability when he began inviting his friends in the spring of 1901. Placed as it was on the side of a rather steep slope, not more than fifty feet from the water's edge, it jutted out from the embankment in some ten or fifteen feet of overhang, held up beneath by slender-looking two-by-four timbers. All around it were the trees, brushing so close that the camp was a veritable arbor, shut in on every side except the frontage that looked out upon the creek.

The extreme front of the large main room, that part of the shack-like structure that overhung the bank, had never been put to the test to determine its strength. Steinmetz speculated for several days as to how strong this side of the lodge might be. At length he resolved upon a convincing experiment, quite characteristic of the man in its naïve mingling of care-free merrymaking and unconscious risk to the merrymakers.

He asked a considerable number of his friends among the young fellows at the General Electric works to come out for a house-warming. He also brought out a band of six or eight pieces. The band was placed as far out on the overhanging part of the main room as possible, close by the large, wide windows. A bowl of fragrant punch was also placed in that part of the room, attracting the frequent footsteps of nearly every member of the party. Steinmetz stood by, looking on as the festivities proceeded, watching to see whether the floor would give way or hold.

Throughout the evening the supposedly doubtful wing of the structure, braced only by its rows of two-by-fours, received all the walking about and massing of human weight that any one could desire for a practical test; and the foundation of two-by-fours held fast! After that occasion Steinmetz never doubted that his camp was stout enough for him to live in it.

The official dress at camp for the men was a bathing-suit, particularly when it was a stag-party. If there were ladies among the visitors, then the young engineers appeared in city clothes. Steinmetz himself, however, invariably wore his bathing-suit while at camp, regardless of whom he was entertaining.

Only one beverage was known at these week-end parties. This was a particularly delicious variety of Swedish punch. Likewise there was no variation in the Sunday dinner. The official bill of fare called for beefsteak, cooked to a tender brown. Steinmetz usually acted as the cook.

After the meals, the men who were visiting Steinmetz lined up around the dish-pan, and a simple but inflexible system was followed in handling the after-dinner camp chores. The first fellow cleared the table; the second cleaned the dishes; the next washed the dishes; the fourth dried them and put them in the cupboard.

The official work at the camp, the one enterprise in which everybody was expected to bear a hand, was the building of a dam across the creek below the camp to enable Steinmetz to paddle around in his canoe without getting stuck on the sand-bars. This was the great "engineering" job at Camp Mohawk, as Steinmetz called his lodge.

Viele's Creek in those days was a shallow stream, in dry seasons little more than a succession of pools. Later it was filled deep by the back-wash that followed the deepening of the Mohawk to form part of the New York State Barge-canal. But in the early years of the Steinmetz camp, there was seldom enough water to allow a canoeist to get very far.

The dam was located at a point where the water as it backed up formed a broad and fairly deep lagoon. It made a fine swimming-pool and pond for canoe-ing. The dam itself was simply a stone wall thrown across the course of the stream and cemented together with gravel.

The whole crowd of camp visitors, led by the enthusiastic Steinmetz, would go out and work on this dam Saturday afternoons and again on Sundays. The work was kept up in spasmodic fashion all summer and continued two or three summers after that. Even then it could hardly be said to have been completed, because the current frequently washed out the gravel cement and loosened or carried away some of the stones.

On Sundays a party of women sometimes went out to the camp, escorted by Mrs. Krueger, who generally acted as camp hostess, assisted by her daughter Gretchen. Steinmetz provided the transportation, in those days consisting of horse-drawn stages, such as were popular for straw-rides in the winter. It was the regular thing on Saturdays and Sundays to hire one or two of these vehicles from livery-stables, while frequently some of the fellows would paddle out from Schenectady in canoes, coming as far up the creek as beyond the dam.

Gradually young Hayden, who was considerably Steinmetz's junior, began to draw into more intimate friendship with the whimsical mathematician and lover of river life. It was not a great while before Hayden seemed to be more of a real companion to Steinmetz than any of the others, except perhaps Mrs. Krueger's boy Carl, of whom Steinmetz was always very fond.

Perhaps this was because of Hayden's readiness to devote himself more than most of the fellows to helping run the place. This, to be sure, caused the two to be thrown more together than would otherwise have happened. Once or twice Steinmetz wanted some one to stay out over Sunday night with him. As nearly all the other young men were obliged to go to work at seven o'clock, they begged off and went home Sunday nights. But Hayden's work did not start until eight o'clock, and so he said he was willing to stay and help in the work of cleaning up, making everything ready for the next week-end gathering.

Meanwhile these spring months of 1901 were unusually absorbing for Steinmetz because of his home-building venture. This project was undertaken without any notion of what it was to lead to, yet it proved one of the mile-stones of his peculiar, eccentric home life, the future hearthstone around which most of the deep affection of his sensitive nature was to be lavished.

CHAPTER XV

IN the early spring of 1901 Steinmetz prepared to build a house of his own, and it seemed to him to be the most tremendous enterprise he had ever undertaken. It absorbed his thought and energy exclusive of everything else, for the time being. He studied over it as intensely as he did over his most astounding mathematical investigations, with all their significance for the welfare of mankind's practical life.

There was this difference, that, whereas he completely mastered the most forbidding equations in mathematics, the matter of building a house was almost too much for him. Problems and formulæ, with all their fearful array of signs and symbols, had no terrors for him; but the necessary details involved in erecting a dwelling, these, indeed, were a mystery of large proportions to the versatile Steinmetz mind.

There were unusual factors, too, that arose from his personal preferences and the aspirations that were dearest to his heart. For Steinmetz knew quite clearly what sort of a place he wanted. He had made up his mind how he would begin to build, what he would do first of all. And it was not the house itself which he put foremost on the list. It was the matter of a conservatory for his plants and pets, and of a laboratory for his electrical and chemical experiments.

Thoughts of his personal comfort apparently were secondary with him. He planned to rough it by living in the laboratory if necessary. But his orchids, his cacti, and his ferns—these must be given shelter before anything else was done! He could not feel easy until he knew they were completely protected against the harshness of the next winter.

Similarly Steinmetz put his scientific enterprises ahead of his home life. Really to live, according to his way of thinking, one must have a well-equipped laboratory, and freedom to spend about two thirds of every twenty-four hours conducting those fascinating investigations which he seemed to enjoy more than fine clothes, good things to eat, or a costly home. Other folk might revel in luxury or drink their fill of the foaming nectar of pleasure; Steinmetz got his fun out of his laboratory work, with his conservatory of peculiar plants and still more peculiar pets to fall back upon in moments of leisure, and, for real holidays, his rough little camp on an inlet of the Mohawk, which was his unfailing source of absolute recreation.

And so he set about building his own house by first building a house for his horticultural collection; and next by erecting a private laboratory with a most orthodox atmosphere of switches and lamps, batteries and transformers, compounds and tubes and retorts. These represented the delights of the scientist, with which he found pleasure in working, although he himself would never call it by that inaccurate, toil-suggestive term.

To have seen Steinmetz in these spring and summer days of 1901, any discerning observer could have sketched out the real life-interest of the man beyond the possibility of error. Only a zealous disciple of science, only a lover of nature and some of nature's peculiar plant forms, could have builded in the manner that Steinmetz builded on that wooded piece of land with its frontage bordering along what was to be Wendell Avenue, and bounded on one side by a little dell, through which rippled a gurgling brook.

The real-estate transaction through which Steinmetz actually acquired his property was carried out in the autumn of 1900. There is good reason to suppose that Steinmetz hoped to begin building operations early enough the following spring to give him a place of his own before his lease on the Liberty Street house ran out in May. Any such expectations were dashed, however, when the first of May arrived, for at that date nothing whatever had been accomplished on the Wendell Avenue property.

Finding himself thus faced by a sudden dilemma, Steinmetz interviewed his landlord and persuaded the latter to let him stay two months longer on Liberty Street. Then he set about to get his new

The original camp of Steinmetz on Viele's Creek, a tributary of the Mohawk River

laboratory and his conservatory under construction with all possible speed.

He made arrangements through E. W. Rice, Jr., then vice-president of the General Electric Company in charge of technical work, whereby the laboratory was built for him by the company at a moderate cost. The conservatory he built with his own funds, as also, later on, the house.

Fairly early that summer both laboratory and conservatory were ready. They were connected by a passage through a basement structure, which served later as the foundation for the house. The laboratory as originally built—it was enlarged in later years —was a square-shaped structure, two stories in height and sufficiently spacious to house all the equipment and supplies which Steinmetz wanted to put in it.

On the ground floor there were three rooms, all quite bare. The walls were brick and plaster, unadorned; the floors consisted of stout wooden planks. The center room was provided with a work-bench along the whole of one side. Up-stairs there were two smaller rooms.

No sooner was the laboratory completed than Steinmetz took possession. Almost immediately he began moving in various electrical paraphernalia and special laboratory equipment. Also, from the Liberty Street lodgings, he brought up a modest array of house furnishings which he had accumulated during the preceding six or seven years. Quite unconcerned as to what sort of home surroundings he might expect, he calmly planned to combine his own living-quarters with the laboratory for as long a period as need be. Meanwhile, ignoring this rather barren prospect regarding his personal comfort, he gave careful attention to the completion of the conservatory, and to the important task of moving his collection of plants into their new quarters.

Queer as the Steinmetz laboratory-home undoubtedly was, it yet breathed the personality of its occupant. The intimates of Steinmetz, walking through the rooms a month after their owner established himself, might have guessed that this was the Steinmetz haunt supreme. A stranger, however, would have been pardoned for regarding it as a den of incomprehensible perplexity.

The rooms all had plenty of window light, for the building was unobstructed on all four sides until the residence was built abutting it on the east. Bulbous electric lights were hung around or snugly stored away on shelves. Glass-paneled lockers held various odds and ends of laboratory supplies. Strange metallic apparatus stood in helter-skelter groups about the floor—switchboard panels, small transformers, and tables or benches displaying rows upon rows of storage-batteries. Chaotic tangles of wires were in evidence in one or two places, or were to be seen passing and repassing overhead.

In the midst of this incongruous display, one or two parlor chairs sat about; a handsome table, which obviously was no part of a laboratory's fittings, a bookcase, and some ornaments of bric-à-brac heightened the contrast. There was an old horsehair lounge, also, and in one of the smaller rooms a table of some dimensions which later became the dining-table. Meals were cooked in this room over a gas-burner; and Steinmetz himself was cook, butler, and chief dish-washer.

The second floor provided sleeping-quarters for Steinmetz and later for Hayden as well, each occupying one of the two small rooms.

If they were called upon to entertain overnight guests, which was sometimes the case during the experiments with the magnetite arc-lamp, some one slept down-stairs on the old lounge, or on an old cot in the wash-room. This wash-room was provided with a fine enameled bath-tub, and there was a separate hot-water tank and gas-heater. As a rule, however, when the gas-heater was lighted, it had a disconcerting habit of emitting a violent puff of flame, which was sufficient to frighten any one strange to the place, however much he might desire to bathe. Steinmetz and Hayden knew how to manage this apparatus to perfection, but Steinmetz, at least, was sometimes inclined to keep the secret to himself if there was a chance to have a bit of a joke at the expense of a visiting engineer.

As for the conservatory, always a spot of placid delight to this oddly disposed scientist, whose heart was as large as his brain, it was snugly built and well occupied from the first. It was laid out in two sections, entirely inclosed in glass, the smaller section opening out of the larger at about the center. In this lesser room was a small pool, where Steinmetz was in the habit of enjoying a morning dip, although during the latter years of his life he discontinued this practice, allowing water-lilies to take possession of the pool and cover much of its surface with their broad leaves. The path running through both rooms ended in a platform with a low railing overlooking the pool, and here Dr. Steinmetz liked to come in moments of thought, to lean upon the rail and gaze down at the motionless water.

Originally the pool enabled the doctor to realize one of his fond dreams. For several years he had set his heart on having a fish-pond, and the pool in his new conservatory was no sooner finished than he had it supplied with various species of fish and water life. He replenished it from time to time in a thoroughly characteristic fashion. Whenever he had terrapin soup in a restaurant, he would ask the proprietor if there were any of the live terrapins left. If so, he would buy one or two of them on the spot, deposit them in his pocket, and take them home, where they were immediately added to the inhabitants of the fish-pond. The quiet delight which he took in throwing these new-comers into the pool and in watching them make themselves at home, smiling to himself the while, never failed to amuse those who chanced to be looking.

The path through the conservatory consisted of a

rustic walk made by cutting the rough branches of trees to a uniform length and laying them together crosswise upon two narrow planks. It ran all along one side of the conservatory, descending a slight slope from the end nearest Wendell Avenue and branching off to enter the smaller room.

Hanging above it and on both sides, in the room of the pool, were many baskets of orchids, a picture of beauty during flowering time. Steinmetz's love of orchids made this one of his favorite haunts. He gave much time and care to cultivating these plants, expending some of his great reservoir of patience upon them; and in return they developed into a most charming collection, giving him that refinement of pleasure which every lover of flowers understands.

The cacti and giant ferns were gathered in one large bed running from end to end of the main conservatory. In a short time, with additions which he procured after moving to Wendell Avenue, they presented an imposing appearance, especially to one not any too familiar with these desert growths.

There was a legend in Schenectady that the doctor's fondness for cacti extended back to those early days when he was enrolling himself under the Marxian banner; and that the cacti came to mean to him a parallel in plant life of Darwin's theory of the survival of the fittest. Some thought, too, that he felt an odd fraternity for this desert-reared plant, which was obliged to grow to maturity amid unfavorable surroundings, necessitating a physical battle, just as Dr. Steinmetz had faced physical handicaps in his own life.

Whatever the secret source of his interest, he eventually brought together a collection of cacti that was said to rank second only to the collection in the Kew Gardens in England. The specimens included scores and even hundreds of the ungainly euphorbia, the old-man cactus, the fish-hook and hedgehog species, the aloe, the agave, and the columnar cactus.

Viewed from the top of the pathway, they made a rather weird picture of tall, thick stalks, outreaching limbs and odd-shaped foliage spreading to the eye at various heights. Some of the giant columnar cacti towered almost to the roof and exhibited queer curling tendrils from their otherwise smooth stalks; others showed clusters of typical thorn-studded, petal-like foliage; and one or two varieties of tropical fruit plants overarched the spectator with great leaves, large and wide, like a green flag held aloft. There was a really bewildering profusion of green things, ferns in thick clusters, and bushes that held out unfamiliar little rough branches, like arms with prickly fingers. Yet they were all hardy and splendid specimens of their kind and reflected the untiring care that was bestowed upon them.

Surely, the visitor was wont to remark to himself, he who gathered such representatives of nature's handiwork as these, typical of the beautiful and the weird, was no ordinary scientist, with blue-steel mind, hard and cold to the stirrings of the heart.

STEINMETZ INVENTS ARC-LAMP

He who was so passionately fond of plant life and of blossoms was fully alive to the real spiritual values in the world about him, and unceasingly developed those elements, side by side with his brilliant scientific advancement.

About the time that Steinmetz planned to build his laboratory and conservatory, a new field of technical endeavor was brought to his notice. The General Electric works at West Lynn, Massachusetts, had become the headquarters of the company's engineering and shop work in the production of street-lighting apparatus. The equipment used in the closed-arc systems, then in common use, was there designed and manufactured under the supervision of H. W. Hillman.

Mr. Hillman, seeking to bring out a more efficient type of street-lamp, introduced the subject to Steinmetz; and he had little difficulty in securing the latter's interest, partly, it appears, because Steinmetz was no longer quite so much taken up with the task of acquainting the General Electric designing engineers with the mysteries of the symbolic method of alternating current calculation.

Mr. Hillman and Steinmetz had several conversations during which the closed-arc system was discussed in all its aspects. Steinmetz, however, began to intimate that he thought a better light-giving material than carbon, which was then employed in these lamps, could be discovered.

CHARLES PROTEUS STEINMETZ

Forthwith he set about to find this better material; and after some preliminary experiments he came upon a substance which he believed was just what was needed. This element was magnetite, which is a higher oxide of iron. Its use in an arc-lamp had never before been attempted; and there were several characteristics of the lamp as finally evolved that stamped it as very much of a technical innovation. The light given forth was found to be a brilliant bluish illumination, which was produced at considerably higher efficiency than was possible with the carbon arc-lamp.

The principal difference, from a strictly electrical point of view, was that, whereas the carbon arc-lamp operated on an alternating current circuit, the magnetite arc-lamp could only be used on a direct current circuit.

After making his own initial experiments and working out his first lamp in fairly practical form, Steinmetz secured the coöperation of the Research Department of the General Electric Company. In Building 19 of the Schenectady works (then located toward the end of the factory yard), a great deal of work was done on the new street-lamp. Throughout the year 1900 this work was continued, and Steinmetz spent a great deal of time in his own laboratory in perfecting the new lamp.

But the magnetite arc did not come into general popularity immediately, because it required direct current. One of the means taken to make its merits known was the proposal of the engineers that a trial installation be understaken; and Steinmetz suggested that this be done in the newly developed Wendell Avenue district of Schenectady. He conferred with General Electric officials about it and secured their approval. Then he had a small power-house built on his own property and installed there a Brush machine, a machine which supplied current to arc-lamps on direct current circuits. The Brush machine in the Steinmetz power-house would supply four amperes of current at three thousand volts on a direct current arc-lighting circuit.

The new lamps were then put up on the Steinmetz property, extending from that spot out through several adjacent streets. There were twenty-five lamps in the installation, each mounted on a tall pole. When they were turned on for the first time there was a ceremonious gathering of townsfolk, engineers, officials of the company, and city officials. Steinmetz himself threw the switch that made the scattered lights flash out luminously in the darkness and brought forth a round of applause from the onlookers.

Steinmetz next conceived the idea of a mercury-lamp to illuminate gardens and shrubbery, as the magnetite arc-lamp illuminated the streets. The mercury-lamp was accordingly brought forth; it was also designed to operate on direct current circuits, in series with the magnetite-lamps. Some of these mercury-lamps were installed on the Steinmetz property, where they were turned upon the foliage of the trees. Others were placed in some of the city parks.

After a year or two Steinmetz came forward with still another suggestion, which he proceeded to work out in his home laboratory. This proposed to do away with the more or less troublesome Brush machines and to run the magnetite arc-lamps without a rotating-machine. In the course of his experiments to this end, he observed that mercury would rectify an electric current from alternating to direct. The outcome of this discovery was the development of a mercury rectifier, which eliminated the Brush machines entirely.

This rectifier, which also proved of great use in the charging of storage-batteries, was made for constant potential low voltage and high current, which was just what was needed for battery-charging.

Soon after this, Steinmetz concluded his experiments in the field of street-lighting and took up other matters; but the introduction of the magnetite arc-lamp, and also of the mercury-lamp, continued, through the organized activities of street-lighting salesmen, until hundreds of thousands of these lamps were installed in cities and towns all over the United States.

A few years later, in 1908, Dr. Steinmetz and some of his associates in the General Electric Company were made recipients of the Franklin Institute's certificate of merit, in recognition of this work on the magnetic arc-lamp. The award was conferred through the committee of sciences and arts of the institute.

Before he discontinued his street-lighting researches, Steinmetz accumulated, by his own laboratory work or that of the company, a complete set of the different models made for both the magnetite arc-lamp and the mercury-rectifier. These were ranged in one of the rooms of his laboratory, occupying all of one side of the room. They made a display, in chronological sequence, from the first crude model down through various modified developments to the final successful lamp, and likewise with the rectifier.

To the lay eye, there would not appear to be much physical difference between these successive steps as exemplified by the models; but each of them involved some technical variation over the one which preceded it, these changes comprising the improvements that were made, one after another, in the unceasing forward march toward a culmination that would prove useful to the exacting needs of man.

When J. LeRoy Hayden first met Steinmetz, at the Liberty Street house where the latter then lived, the work on the magnetite arc was just under contemplation. Within the next year Hayden left Schenectady and went to Lynn, where he acquired experience in street-lighting engineering work, just as the magnetite arc-lamp was coming into prominence. A year later he went back to Schenectady to enter the testing-room.

This was about the time that the public display of magnetite arc-lamps was undergoing installation at Steinmetz's grounds on Wendell Avenue and in the adjoining region of the city. From the beginning it was something of a problem to find a man to operate the Brush machine that supplied the lights with direct current. Hayden had become familiar with these machines while in Lynn, which resulted eventually in his selection to supervise the one at the Steinmetz power-house.

It was largely night-work, and quitting-time did not come until a pretty late hour. Steinmetz, noticing this, proposed, after a few nights, that Hayden should sleep at the laboratory, instead of taking half or three quarters of an hour to get back to his lodgings. He offered Hayden the use of the other room on the second floor of the laboratory, and in addition a place at his dinner-table, which meant a daily sample of his own cooking.

Hayden promptly accepted; and thereupon began the association of these two men, one of them so much the elder of the other that the latter found it quite natural, as time went on, to address his companion as "dad." With hardly a break from that day until the end of Steinmetz's brilliant life, he and Hayden were intimates to a degree that transcended the ties of mere friendship and led finally to the legal adoption of Hayden as the foster-son of Steinmetz.

STEINMETZ INVENTS ARC-LAMP

Their life together at the laboratory on Wendell Avenue lasted from the latter part of 1901 until the spring of 1903, when Hayden married and went off to live for a few months in his own home, only to return in the autumn, bringing his wife with him, to establish the permanent Steinmetz-Hayden fireside in the Steinmetz residence, where they all lived together as one family—and a very happy family, into the bargain.

In the words of Mr. Hayden, they "had a free-and-easy life of it in the laboratory." "Dr. Steinmetz," says Hayden, "was forever busy with his experiments. He spent only a few hours at his office at the General Electric works; but he put in a great many hours in that laboratory of his. Sometimes he was out there at all hours of the night. As time went on, and his definite laboratory undertakings became somewhat fewer, the laboratory became a great experimental shop for him. If he read about any new theory in electrophysics, in electrical engineering, or in chemistry, out he would go to the laboratory and tinker around there, trying to work it out for himself. He always wanted to try an idea himself, in the form of an actual experiment; and no matter if some one else had done it, he wanted to do the same thing himself whenever he read about other people's experiments."

The life of Steinmetz and Hayden there in the laboratory was much less domestic in its aspects than almost any other period of Steinmetz's whole career.

It was more haphazard, so to speak, than when Steinmetz roomed with Asmussen in New York, because now everything was subordinated to laboratory enterprises; it was less systematic than the life at Liberty Hall, because there was no housekeeper to call the men to meals at regular hours and to give them a well-balanced diet.

Steinmetz cooked for them both. And he did it well, for, being a chemist, he understood cooking to perfection. But the actual operation of preparing a meal was, with him, simply a necessary part of the conduct of the laboratory. He viewed it as one more scientific process, incidental, of course, to the main undertaking, but none the less part of the same general program.

His favorite meal was beefsteak and boiled potatoes. Other vegetables were rarely to be seen on the table. These they had day after day, with only an occasional variation. And all other items on the menu were governed by the reaction of Steinmetz, the chemist. They were selected, not by any laws of nutrition or the calculation of calories, but by a chemical color standard. And the chemical color to which Steinmetz was peculiarly inclined was a deep yellow. Foods that approached this tint, when cooked, were his unfailing choice—whenever he cooked anything besides the inevitable beefsteak and potatoes! Egg dishes, consequently, came to occupy a high place on the list; and, as time went on, Steinmetz developed a pronounced taste for flapjacks and griddle-cakes, as well as rare skill in their preparation and cooking.

In the midst of this singular mode of living, in which the usual domestic routine was strictly subordinated to intense mental activity and ceaseless scientific work in the fascinating laboratory, where mysterious blue lights burned every night, and men came and went in a steady hum of activity centering around the quick-moving, quick-speaking, yet cheery little giant of electrical science—in the midst of all this, there came to Steinmetz the first of several honors which indicated the esteem in which he was held by other scientific and learned men. It was his election in 1901 as president of the American Institute of Electrical Engineers, to serve for the year that ended in June, 1902.

It was just about ten years since Steinmetz had stirred the members of the American Institute of Electrical Engineers by the first of his epochal papers on hysteresis, delivered before that distinguished body on January 19, 1892. Then he had been quite unknown except to a very small group. He was merely an obscure young engineer, hardly more than a draftsman in actual status, who had only been in the United States a year and a half, and whose English was, as yet, far from fluent.

The swift span of one short decade had worked a marvelous change in his professional position; a change so extraordinary, so dramatic, that the American Institute of Electrical Engineers did not hesitate now, in 1901, to honor the man to whom it had listened in surprise in 1892, by elevating him to its highest office.

A year later, in June, 1902, Steinmetz was invited to the commencement of Harvard University to receive the degree of master of arts. In conferring this degree, Dr. Charles W. Eliot, president of Harvard, made this memorable statement:

"I confer this degree upon you as the foremost electrical engineer in the United States, and therefore the world."

And here was another dramatic episode in the career of this foremost electrical engineer. In his native country he had just missed winning his scholastic degree at the University of Breslau, although the reason for this lay in his own policy of espousal of democratic socialism. Yet from the collegiate point of view he had always been entitled to the degree; his original thesis, published shortly after he came to America, showed the thoroughness of the work which he did to qualify for such a distinction. And now, with the added brilliancy of his practical work in mathematical research, engineering developments, and invention, the universities of his adopted country felt that the honor should not be withheld.

In keeping with this attitude, the following year brought him still another recognition of this sort. Union College, in his home city of Schenectady, conferred upon him the degree of doctor of philosophy and invited him to become a member of its faculty as professor of electrical engineering.

Steinmetz was now, therefore, a doctor, the title thenceforth being everywhere coupled to his name. He bore this recognition quietly but worthily, and continued to embody, in achievements and personality, the typical doctor of scientific learning.

CHAPTER XVI

FORTY years of engineering progress were sufficient for men to conquer the problem of transmitting electric current across long stretches of territory. In 1880 Edison sent current for his newly invented electric incandescent lamp about a mile at 220 volts. For a longer distance it was found that heavy losses of energy occurred, even if only a small amount of current was transmitted.

In 1922 there was put into operation a transmission-line which flashed electrical energy at 220,000 volts for 250 miles. It is capable of sending this enormous volume of energy even 270 miles.

Fifty thousand horse-power energy may be "stepped up" or "stepped down" in the process of transmission with a loss of only seventy-five one-hundredths of one per cent for each operation.

The name of Steinmetz is inseparably linked with this triumphant development. Others, of course, did their share; but it was Steinmetz who solved some of the fundamental problems. And he was thus engaged during this epoch, embracing the years from 1900 to 1910. In his later contribution to the general problem his ability as an inventor came into evidence, supplementing his earlier basic work in engineering mathematics.

But his immediate personal concern in the years now under consideration was in the building of his residence on Wendell Avenue. Begun late in 1901, this comfortable house was not completed until well into the summer of 1903. It was built in a peculiar fashion; it might almost be said to have been built piecemeal.

As finally completed, the house presents an imposing example of architecture of the Elizabethan period. It is built of ornamental red brick, with high-peaked gables and a substantial dark wood finish, and has commodious, attractive rooms. The general plan is in keeping with the suggestions of Dr. Steinmetz and embraces a large, hospitable entrance-hall, broad stairs with heavy, wide banisters, and certain touches particularly adapted to the convenience of the owner.

Chief of these is an office-room and museum, opening out of the entrance-hall at the extreme rear, and reached through an anteroom. There Dr. Steinmetz had his desk, on one side of which was a large swivel arm-chair, for the special accommodation of visitors; while on the other side stood a low bench, suggesting a leg-rest, with a cushion, on which the doctor placed himself, knees on the cushion, elbows on the desk— his favorite attitude when in conversation with any one who came to see him. At one elbow stood a large bowl, the cover of which he would raise after a few moments, to reach in and take a long, thin cigar, which he would slowly light and contentedly smoke throughout the discussion.

This room was large; and more than half the space was occupied by half a dozen glass-inclosed shelves, which stood in rows with aisles between them, and which reached clear to the ceiling. These contained eventually a miscellaneous assortment of odd treasures, the hobbies of a scientific collector extending over the better part of a lifetime.

In a corner of one case were incandescent lamps of all sizes, even up to the largest. On another shelf were to be seen tray upon tray of geological specimens, iron ores and minerals of all sorts. Another contained a large display of Indian flint arrow-heads, including some excellent specimens. There were also collections of curious smooth stones and pebbles, sea-shells, old note-books, hour-glasses, and many other objects. There was a ladder, on wheels, which the doctor used to get at the upper shelves of this museum.

The office opened, at its far end, directly into the laboratory; and the house was also connected with the conservatory by a door opening off the main entrance-hall. This fulfilled Steinmetz's original plan of having the house fairly well toward the front of the property, with the laboratory at the rear and the conservatory on one side.

The building of the house began with the dining-room. This room was virtually finished as a unit before any other building operations were undertaken; and it was so located that it overlooked the pool in the conservatory.

When, little by little, the rest of the house finally began to take shape, the dining-room constituted the nucleus of the whole. Around it the house proper was laid out. But the work was not a continuous process. After the building of the dining-room there was an interval when nothing was done. Apparently Dr. Steinmetz employed more than one building contractor or was undecided in his own mind about some details of the plans.

Hayden, who was then with him, and who watched proceedings throughout 1902 and into 1903, particularly noticed that Dr. Steinmetz was not easy in his mind about the house, even though he saw it finally taking shape. For one thing, Steinmetz had no particular plans for living in the residence after it was finished.

He told Hayden several times that he did not know what in the world he was going to do with the house after it was built. It was entirely too large for him to live in alone; moreover, he had no furniture and had no plans for acquiring any until he knew how he was going to use the house.

Apparently he never gave a moment's thought to the various matters involved in maintaining the house. And as he saw the time drawing nearer and nearer when this problem would have to be worked out, he began to feel dismayed.

The work continued, however, and in the autumn of 1903 the house was finally turned over to its owner. Around it there was—and still is—considerable ground. The passer-by in these present times is wont to notice the wide expanse of green lawn, with its hedges and graveled walks. But it displayed none of this when Steinmetz first took possession. The lawn not only did not exist then but was not even thought of. Steinmetz did not want a lawn; did not, for some queer reason, like to see lawns. Consequently, for the first year or two, the house was surrounded by a field of tall, uncut clover, which nodded in the summer breezes and grew up around the home until the spacious residence seemed fairly lost in it. After the Haydens took up their abode with him, the doctor was persuaded to allow a lawn to be laid out, and the odd-looking landscape round about thereupon became more conventionalized.

Mr. Hayden married in the spring of 1903, thereby relinquishing his quarters in the Steinmetz laboratory. Steinmetz witnessed his departure with considerable regret, which he did not hesitate to express.

It is obvious that he had formed a genuine attachment for the younger man, and that the prospect of reverting to a solitary life was not at all to his taste. Above all, he appears to have shrunk from living alone in his big new house. Withal, his big, friendly heart yearned after Hayden as he watched the prospective bridegroom pack up and go his way.

The suggestion of pathos at this little crisis in Dr.

PROFESSOR AND ENGINEER

Steinmetz's life is unmistakable. He has the appearance of a lonely figure, an effect which is heightened by the background of his comfortable material environment and his fine new house, as pretentious and attractive as any dwelling in town, yet unalluring to him without the prospect of human society.

Hayden was married on a Friday in May, 1903; the couple went on a very short trip, returning to Schenectady on Sunday, and setting up housekeeping in a flat on Park Avenue. On Monday night they were agreeably surprised when the door-bell rang and Dr. Steinmetz appeared.

"I wanted to see you again; I just came over for a little visit," he explained, smiling ingenuously.

He was immediately invited to supper; and the three had a pleasant evening together, Dr. Steinmetz finding evident enjoyment in discovering that his old laboratory mate had not lost any of his congenial traits through the acquirement of a wife. Mrs. Hayden, incidentally, was no stranger to Dr. Steinmetz. On several occasions before her marriage she was one of a party of visitors at Camp Mohawk, on days when the camp and its genial proprietor were entertaining the ladies.

Throughout the summer of 1903 Dr. Steinmetz was a frequent visitor at the Hayden home. He never failed to report on the progress on his house, then in its last stages of construction. And when, in the early autumn, it was all ready for occupancy, he invited Mr. and Mrs. Hayden over to see it. As they went through the empty rooms, remarking upon the handsome interior finish and the snug, substantial aspect of the work, Dr. Steinmetz suddenly exclaimed:

"Why don't you both come here to live? We will all have a home here together. It will be pleasant for every one."

And so it came about. Both Mr. and Mrs. Hayden were attracted by the house, and both thought a great deal of Dr. Steinmetz; while the doctor, on his part, was as pleased as a boy when they accepted the proposition. The next few weeks were busy but happy, for him, as he saw his friends take possession, and the empty rooms take on the appearance of a real home.

Whenever he felt really jovial, as he did in these autumn days of 1903, he was given to indulging in various whimsical, boyish pranks; and he and Hayden treated Mrs. Hayden to an amusing surprise, out in the laboratory, during the festivities incident to the settling of the home. The surprise was a box of candy that had been subjected to an electrical charge. When Mrs. Hayden essayed to take it up she received a sharp sensation that made her fingers tingle and caused her to drop the box with a startled exclamation, while Dr. Steinmetz, looking on with Hayden, chuckled in quiet amusement.

Jokes like this were a favorite of the doctor's for a while. It was the sequel to his acquirement of a static machine, which some friend had given him.

PROFESSOR AND ENGINEER

The machine stood in one of the rooms of the laboratory, where it was of no use whatever to Dr. Steinmetz in his work. But, being a capital means of charging a table or a chair or any other article with electricity, it came to be valued by him simply for its fun-provoking possibilities.

Thus it was sometimes used by Dr. Steinmetz, with the connivance and assistance of Hayden; until visitors to the laboratory would hardly venture over the threshold, and could not by any means be induced to sit down or to take anything in their hands if they did enter.

Fortunately Dr. Steinmetz was careful to play tricks with this machine only upon his very closest friends, people who would enjoy the joke as much as he did. When either Hayden or he saw such a friend approaching the laboratory, they would hasten to the static machine and turn on the current. Then, when the visitor touched a table, or sat down in a chair, he received a vigorous shock. Perhaps when he even shook hands with Dr. Steinmetz, if the latter brought his other hand into contact with an object that was charged, the new arrival was instantly electrified, in a very literal sense, by the greeting he received.

In the late winter of 1902–03 Dr. Steinmetz entered upon his duties as professor of electrical engineering at Union College. It was the beginning of a ten-year period in his life devoted to a unique leadership among young men, a period which left its impress upon the slowly shaping destinies of numerous fresh young lives and simultaneously stirred in Dr. Steinmetz's own life his ever-sensitive social instincts, appealing directly to the fraternal side of his nature.

The circumstances that led to his joining the college faculty arose principally from the close working interest existing between Union College and the General Electric Company. The company was always friendly toward the college; if the need arose, it stood ready to lend a helping hand.

In this instance, the helping hand was a boon indeed. It was extended at the request of Dr. Andrew Van Vranken Raymond, then president of the college; and it brought to the institution the man of whom it was said around the campus after his death that he was "the most unique as well as the most widely known professor ever a member of the Union College faculty."

Previous to this time, Union College had offered merely a nominal course in electrical engineering, conducted by Assistant Professor Horace T. Eddy. The acquisition of Dr. Steinmetz infused a tremendous stimulus into this course. It was no longer a subordinate branch of the general curriculum but became an established department of the School of Engineering.

The catalogue of the college for 1903–04 describes the new department in words that clearly reveal the influence of the Steinmetz ideal of a well-balanced training, combining both the classical and the technical, for men who would become electrical engineers. The department of electrical engineering, according to the catalogue, "aims at a thorough and broad scientific education of the prospective engineer, rather than the specific training of a specialist." Then are named the three classes of studies which constitute "such training as is now considered essential for every educated man": first, a general culture course; second, languages, literature, and history; and third, "a broad and general technical education," especially striving for a thorough understanding of the fundamental principles, "rather than a memorizing of numerous facts—aims at quality, and not quantity."

President Raymond made the first definite public announcement of the arrangement whereby Dr. Steinmetz joined the faculty of the college at the fifteenth annual banquet and reunion of the Union College Alumni Association of New York, in December, 1902. In his remarks on this occasion he credited the General Electric Company with a spirit of friendly helpfulness through which "the special electrical expert of the company, whose position at the head of his profession is unchallenged, the man to whom President Eliot referred last June, when conferring upon him an honorary degree, as 'the foremost representative of electrical science in America, and therefore in the world,' Mr. Charles P. Steinmetz, is permitted to take charge of our course of instruction, and has already been appointed professor of electrical engineering in Union College."

Dr. Steinmetz was present at this gathering; and on being called upon to respond to the toast "Lightning Progress," he gave an outline of the electrical engineering course which he conceived as appropriate for Union College. He seized the opportunity to bring forward his belief in a "broad and general scientific education, in all the knowledge required by the educated man in this twentieth century," to quote from the college catalogue.

The "Concordiensis," published by and for the students of the college, contains several references to Dr. Steinmetz and the new department during the next few issues of the publication. In the number for February 18, 1903, Dr. Steinmetz himself has an article, discussing his plans and his theories as to what constitutes a proper course of study for a man who wants to become an electrical engineer. On March 11, the "Concordiensis" reprinted an article from "Success," in which Dr. Steinmetz told in his own words the story of his boyhood and student days in Breslau.

Dr. Steinmetz served the college without financial compensation. Never a seeker of wealth for wealth's sake, in the case of his professorship he held the quaint, perhaps old-fashioned, German conception of the teacher. In his native country it was always looked upon as a most honorable profession; the "Herr Professor" was regarded as a personage set

apart. Dr. Steinmetz carried this point of view with
him when he accepted the Union College professorship. He clearly considered that he was honoring
himself in assuming the position.

His work at the college gave added importance to
the laboratory of electrical engineering which was
established on the campus during this period, and in
which E. E. F. Creighton was a prime mover. Interesting work was done there, especially in the study
of corona, or electric current leakage during transmission. A miniature transmission-line, built with
great care to a scale of nearly two thousand miles,
was part of the laboratory equipment devised by Dr.
Steinmetz and Mr. Creighton.

As professor of electrical engineering, Dr. Steinmetz was the head of this entire department at the
college. He planned the curriculum of the department, engaged those who served under him, and directed their work. He was in every sense an active
professor, lectured regularly every day to seniors and
juniors, and attended all the faculty meetings he
could find time for, incidentally adhering complacently to his fondness for smoking there as at the
General Electric Company. He was the only professor who was ever known to smoke during a faculty
meeting at Union.

In this new realm of activity, Dr. Steinmetz found
a unique opportunity to exercise his talents as a
teacher; and to the extent of those talents a great

many men who came under his instruction in the decade that followed will readily testify. He made them feel at ease, bore with them in their perplexities, and displayed such a personal interest in the individual as forever removed him from the proverbial classification of "cold, calculating scientist."

He could hardly be called a strict disciplinarian in the class-room. Yet his lectures were usually so well worth while, so original, and so lucid as to hold the interest of his students through an entire course. If any situation arose in which he was moved to comment on some laxity of procedure in his classes, he invariably did so in characteristic quiet fashion. His manner was mild; but it was always evident that he was perfectly aware of what was going on, knew instantly when he was taken advantage of by the thoughtless or the drone, and was too discerning to be deceived, no matter what the occasion.

From the very first, he was popular with the students. Although maintaining a certain reserve, which was caused by shyness much more than by assumed dignity of any sort, he was yet fraternal with the students, joined the Phi Gamma Delta fraternity, and aided the boys of that organization in securing a fraternity house. He was occasionally a visitor at the Phi Gamma Delta house himself, mingling in a friendly, quiet way with the students, all of whom looked up to him with admiration and respect.

He later became a member of Sigma Xi and Tau

Beta Pi, honorary scientific fraternities; and of Eta Kappa Nu, engineering fraternity.

Those were the days before the institution of a compulsory athletic tax upon the student body. The only method of supporting college athletics was to canvas for funds. And when the canvassers went their rounds they could always count on Dr. Steinmetz for a substantial contribution. He was keenly interested in their sports, as in all outdoor life; had tried several times to learn to swim; and was always ready to help along any athletic enterprise among the boys of the college.

The college Y. M. C. A. frequently secured Dr. Steinmetz as a speaker on popular subjects at Sunday afternoon social gatherings. These were always the most successful of the Y. M. C. A. events. The college students thronged to hear their idolized professor of electrical engineering. Sometimes he spoke of his engineering experiences, or recounted days in Germany when he was himself a student. But his personality and appreciation of humor, as well as his faculty for explaining and describing, made these talks of unfailing interest to the students.

The first time that Dr. Steinmetz attended a student play at the old Van Curler Opera-house, he caused a thrill among the student body by appearing in one of the boxes attired in a gray flannel shirt beneath an ordinary sack-coat. The students were greatly fascinated by this. It struck them as absolutely characteristic. Nobody but "Steiny" would have done it; and nobody but he would have done it as a perfectly natural, ordinary practice.

"In short," as the "Union Alumni Monthly" puts it, "the boys took him to their heart, and he entered fully into the life of the college."

In the lecture-room he was inclined to set a pretty stiff pace for his classes. He was wonderfully stimulating as a lecturer, but he frequently talked far over the heads of the students. The reason for this was twofold. He never quite realized the limitations of the student mind. With all his sympathy and kindly interest in individual members of his classes, he always imagined that young men could absorb more than was to be expected. In addition to this, he was apt to feel stirred to a high pitch of enthusiasm within himself as he got up into the upper levels of the grand field of electrical engineering, and so would unconsciously leave his admiring, but bewildered, classes far behind.

Nevertheless, he was a successful class-room lecturer and exceedingly popular with his students. At first the college boys found his high-pitched voice and his marked accent a trifle difficult to follow, but this reaction did not last. He possessed a remarkably clear diction; and he invariably unfolded his subject in a spirited and entertaining manner, so that even a technical lecture, as he handled it, was decidedly effective.

As a result, his classes gained a vivid impression of the broad background of the subject, securing

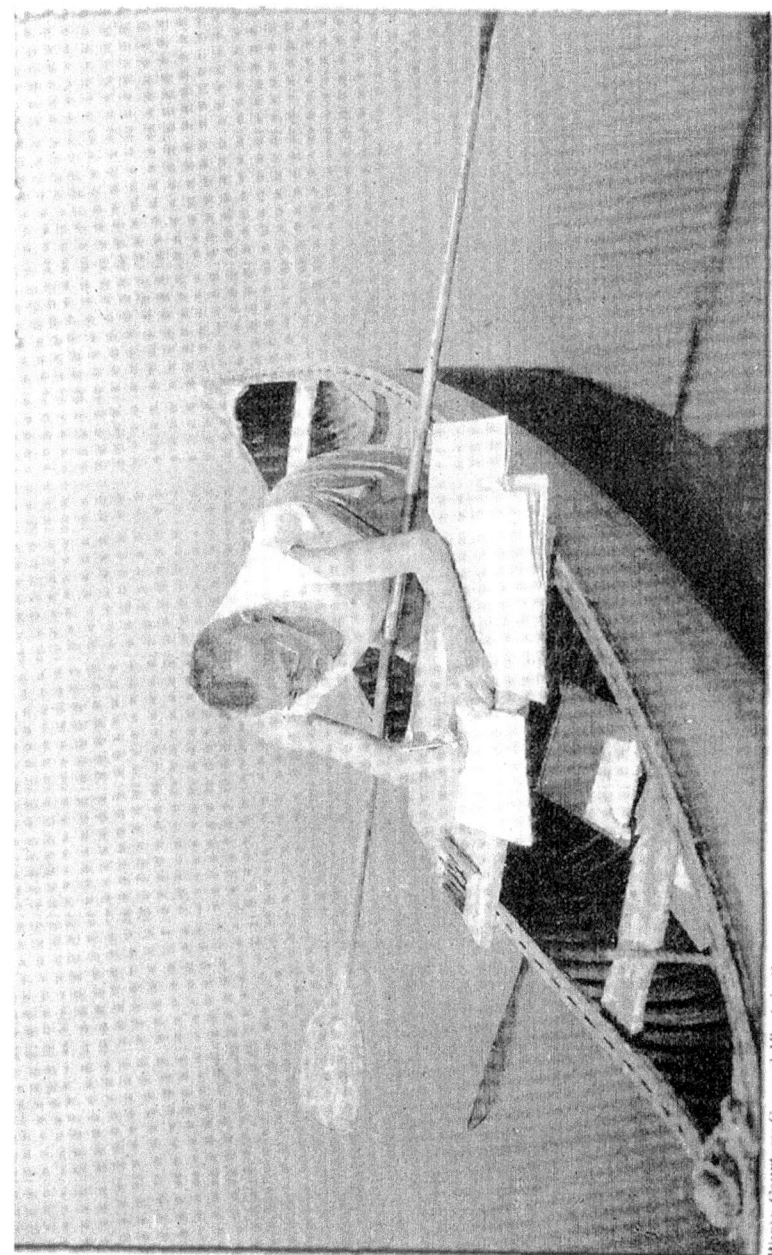

Dr. Steinmetz at work in his canoe at the camp, during the summer of 1923. In this little craft he would float up and down the creek, figuring out intricate problems in electrical engineering

a sound conception of the fundamentals. Even though sometimes left floundering by the ultra-technical exposition to which they had listened, the students of Dr. Steinmetz carried away with them a realization that electrical engineering was a vast and tremendous field, a profession of unimaginable possibilities, in which no man with ambition and opportunity could feel restricted.

Soon this came to establish a reputation for the department of electrical engineering at Union College. Other educational institutions realized that under Dr. Steinmetz the students at Union were getting a conception of fundamentals which was bound to fire them to unusual endeavors. Dr. Steinmetz's contagious faith in his subject, an overpowering personal enthusiasm, and a manner of presentation that was fascinating to all except the utterly indifferent were responsible for this result, almost as much as his utter mastery of the course.

Thus it came about that the boys who got the most out of Dr. Steinmetz's classes got it through the medium of inspiration. There was awakened within them a feeling of respect for electrical science, a perception of what a mighty thing it could become in the shaping of the world—exactly the feeling that Dr. Steinmetz himself had about it. His soarings into the glorified, cloud-enshrouded upper heights, where he left all others behind him, had the effect, at least, of revealing to the students the boundless magnitude of the field, leaving them murmuring to themselves, "What a tremendous thing is this of electrical engineering!"

Yet that one failing of Steinmetz persisted: he kept forgetting that other people, especially students young in years and experience, could not grasp mathematics and science as easily as he could. Although he had the ability to make things clear, although he possessed excellent didactic powers, although he would start his lectures well within the understanding of his classes, he was almost sure, as he went on lecturing, to leap to limits which they could not reach.

Dr. Steinmetz lost no opportunity, while he was a college professor, of urging his theory of a broad general education for technical men. He always insisted that no student could fully understand electrical engineering in just four years. That, he contended, was merely long enough to enable him to lay a foundation, upon which he must perseveringly build as time went on.

But a technical training, he likewise maintained, is not a narrow thing, confined largely to one specialized subject. It must be accompanied by a grand study of the classics, of Greek, of psychology, history, economics, and philosophy. Proceeding on this theory, he guided his students toward gaining a wide, sweeping view of the whole physical world and of all that had ever entered into it in times past, pointing out that thus only could a student perceive the significance in the relation of one thing to another.

PROFESSOR AND ENGINEER

As can easily be imagined, this generous educational policy promoted the most harmonious relations between the engineering and the classical faculties of Union College during Dr. Steinmetz's time. The classical department found no friction or cross-pulling from the technical department, but instead a loyal and sincere support of its own curriculum—an outspoken missionary in what is all too frequently looked upon as the competitive camp.

There are a few deliciously typical Steinmetz anecdotes still told about the campus, one of which seems likely to become enshrined as a permanent college tradition. It harks back to 1904, when Dr. Steinmetz had just begun to "let himself out" as a college professor of unmeasured ability. In that year he had one lecture to juniors which lasted two solid hours and usually left the class in a somewhat limp state of mind. In the middle of it he allowed a fifteen-minute recess, during which the boys could move about or leave the room. After resuming, for the second section of the period, the doctor would call the roll again.

It happened one day that nearly half the class lost heart at the end of the first section of the lecture and did not appear when the second roll-call occurred. The absent ones had arranged with those who stayed through to answer for them when the names were called; hence, as the doctor went down the list, a voice responded "Here" for every name, yet, upon glancing up as he finished the roster, the doctor perceived that out of perhaps fifty members only about twenty-five were present.

"Twenty-five men are here," he announced, with a characteristic hitch of his trousers, "yet fifty answered to the roll-call. A remarkable mathematical phenomenon!"

It is also related that a student one day asked Dr. Steinmetz if there is any truth in the old adage that lightning never strikes twice in the same place.

"Yes," he replied, "I may say it is true in a great many instances."

"Why is that?" persisted the student.

"Because," said Dr. Steinmetz, with a twinkle, "there is usually nothing there for the lightning to strike a second time."

In 1913, Dr. Steinmetz expressed a desire to relinquish the duties of an active professor. He realized that his professorship was bound to become more and more of a side issue, and he felt that the subject was too big and too important to be a side issue.

He recommended as his successor his old chum of Liberty Hall days, Dr. E. J. Berg; and, at an informal little dinner given that year to both Dr. Steinmetz and Dr. Berg, the former delivered a short speech which was in the nature of a farewell to his associates on the faculty and a presentation to them of his successor.

In the following year Dr. Steinmetz's name appeared in the Union College catalogue as professor of electrophysics, a position which he retained until his death. But it did not involve very much activity. As a matter of fact, he lectured only once or twice a year, and after 1915 his relations with the faculty were merely nominal, although he remained a friend of the college the rest of his life.

Dr. Steinmetz retained his connection with the General Electric Company throughout his period of service on the college faculty. He seldom spent more than two or three hours at the General Electric works; and his employers allowed him such latitude of procedure that there was no possibility of conflict between his activities at both places.

His work for the General Electric Company had now come to concern itself with special investigations, the problems of the moment, whatever new difficulties arose requiring unusual technical ability to unravel.

Among these was the subject of transmission. More and more, the demand of the time was for long-distance transmission of electrical energy. And, in this field, electrical engineering still had a great deal of research to perform. At the same time, the electrical engineers possessed such vision as enabled them to keep well ahead of existing practice.

They all foresaw, Steinmetz with the rest, that the voltages at which electrical energy was to be transmitted from place to place were bound to undergo a vast increase. As these voltages went up, the problem of finding transmission-cables which would carry such high electrical pressures, and the whole question of practical high-voltage transmission, proved perplexing.

Voltages higher than 75,000 or 80,000—the pressures then in common use—caused corona to appear around the transmission-cables, which indicated that some of the current was leaking because the cable was not of sufficient capacity to carry it. Extremely high voltages also created interference with the operation of electrical apparatus in the power-stations.

Hence the investigation resolved itself into a study of cables and of means of eliminating disturbances in electrical machines. In this work Dr. Steinmetz played a prominent part. He helped to establish an extension training-course, which was intended to supplement the regular test-course at the General Electric works. This extension course sought to train groups of engineers for special lines of engineering work, one of which was the study of transmission problems.

The investigations which Dr. Steinmetz carried on in 1903 and succeeding years involved the building of an experimental high-tension transmission-line in the yard of the General Electric works at Schenectady. A transformer designed to raise line voltages to 220,000 volts, the highest then known even in experimental work, had been built for use with this system —twenty years before such voltages were demanded by commercial usage.

The work had satisfactory results. Aided by Steinmetz's powerful mathematical ability and inventive mind, new types of transmission-cables were developed, new methods of operating transmission systems were revealed, and transformer improvements were discovered. As an outcome of the whole series of experiments, the engineers were ready when a demand came for a transmission-line going higher than 80,000 volts.

It is interesting to observe that while preparations were being made, years later, to establish the first 220,000-volt commercial system, the engineers in their laboratories were experimenting with transmission at a million, and later at two million, volts. The attribute of foresight, as in the case of Steinmetz, makes torch-bearers of the men who are blessed with it and with a knowledge of how to use it.

Steinmetz the inventor was at this time working wonders similar to those previously achieved by Steinmetz the mathematician. He invented a complete method of high-tension transmission which made it possible to perfect the modern far-reaching electrical systems, and to utilize to the full the transformer invented by William Stanley.

This contribution of Steinmetz to transmission practice solved difficulties which no one else seemed able to dissipate. Much of the study which preceded it was done in the power-houses of Chicago, where an incessant demand for improvements, on account of the ever-growing use of electricity by people of that metropolis, brought the situation to a focus.

CHARLES PROTEUS STEINMETZ

Of all the two-hundred-odd patents which developed from Steinmetz's work, in the field of invention, he himself regarded the series covering this subject as the most valuable. In reply to a question, once asked him, as to what he considered his most important invention, he immediately replied:

"I consider my method of electric current transmission as the most important of my inventions."

He had said upon another occasion, also, that the development of an economical method of transporting power—that is, electrical energy—was the principal accomplishment of modern times in the general technical sphere of work.

Dr. Steinmetz was now so much occupied with investigating special problems that he believed there was a need for a consulting department. He had in mind a group of engineers, each somewhat of a specialist in a particular field, who would be available when needed by any of the company's engineering or commercial departments.

The idea had the approval of the officials; and forthwith Dr. Steinmetz organized such a group, which was designated as the consulting engineering department. From the outset, he was the head of this department, which was established in 1909.

Several men who later distinguished themselves individually quickly joined the department. Among them was Frank W. Peek, Jr., who was interested in corona investigation and the study of transmission problems. After a period of association with Dr.

Steinmetz, he went to the high-voltage laboratory of the company at Pittsfield, where in 1922 he was one of the engineers engaged in a demonstration of artificial lighting at two million volts, simulating and even exceeding in some aspects the startling experiments conducted by Dr. Steinmetz the preceding year or two at Schenectady.

As head of the consulting engineering department, Dr. Steinmetz made special consulting work his particular interest. It is not definitely on record, of course, but there is no recollection, so far as can be ascertained, of any serious difficulty encountered by electrical engineers of those days which Steinmetz was unable to solve.

About this time a problem arose in connection with the first use of electricity in the New York subway. It was found, to the dismay of officials of the electric transit lines, that an electric current was able to pass through a concrete structure, and that, because of this, patrons of the subways might at times be placed in danger. Dr. Steinmetz was hurriedly called into consultation. He visited the subway, discovered the reason for the phenomenon, and showed how it could be obviated.

Repeatedly he gave evidence of his uncanny ability in mathematics. His associate engineers always understood that he carried the entire table of logarithms up to 1000 in his mind and that he knew by heart the seven-point extensions of the table. No one could surpass him in rapid calculations in intricate equations, decimals, and fractions; and to multiply one huge number by another was to him but the matter of a moment, something that he performed almost between puffs on his cigar.

He could talk "over the heads" of the most astute technicians of his day, yet could, and often did, explain an engineering situation to a mystified layman so that his listener could understand the fine points.

He began to be called the wizard of Schenectady; but it was a term that he disliked. He regretted to hear any one call him a wizard—perhaps because it implied some sort of grand isolation from his fellowmen. And that, to Steinmetz, with his human yearning for friends and associates, was an intolerable thing.

CHAPTER XVII

TO say that Charles P. Steinmetz was a Socialist is to make only a half-statement. More
than that, it gives a misleading impression to
a great many people whose conception of a Socialist
embodies some sort of rabid social crank; Steinmetz
did not belong in any such category as that.

He was a genuine idealist—an agitator in his early
years, a dreamer of practical dreams in later life, and
something of a seer. Stated in mere matter-of-fact
words, he was an active member of the Socialist party
in America, and affiliated with its local organization
in Schenectady. Beyond this trite statement, another, almost equally trite, may be brought forward:
Dr. Steinmetz belonged to the conservative wing of
the Socialist party.

But nothing in this dry recital of universally
known facts reveals the inner motives of Steinmetz
which had their abode somewhere behind his formal
Socialistic designation. To catch the real feeling
that was early stamped upon his very heart, and that
produced, in modified and somewhat polished form
as he matured, the grand sweep of doctrine which was
to him almost as much a religion as a political creed,

one needs at least a glimpse of the man's soul in his younger years; one needs to see him intimately during the formative age, the period when everything concerning human life as lived in this complex world of ours appeared to him in such burning reality as to cause reactions in a peculiarly sensitive mind that flamed clear down to the roots of his being.

Such a glimpse—although it is nothing more than a glimpse—is fortunately afforded by the recollections of a man who knew Steinmetz in his student days in Germany, when they were both young together; knew him in his moments of youthful political passion, saw and heard him in moods that discovered the unmasked broodings of his ardent mind.

For Steinmetz at the age of twenty-one, when he was going through the University of Breslau and tutoring in his spare time to meet his expenses, was brooding. In his own circle and his own city, he had observed what impressed him as economic injustice. He felt convinced that there was something wrong when some people had things easy in life because of their wealth, while others, who he believed were equally deserving, seemed doomed to a lifetime of hardship through lack of wealth.

Steinmetz himself, as already remarked, found it something of a struggle to get what he most wanted at the moment, an education. Hence the tutoring of gymnasium students, to help finance his university course; hence, also, his prompt response to the invitation of Hinz Lux to join the university Socialist group, where he worked with an aggressive hostility of feeling that he completely lost in his later years in America, after he had adopted the methods suggested by mature reflection—and had found, incidentally, that Socialists in America were not persecuted as they were in Bismarck's Germany.

Among his pupils, in those days when he was both learning and teaching, was a young man named George Moser, also a native of Breslau. It was in 1886, when Steinmetz was at the height of his student Socialist activities. Moser was preparing to graduate from the gymnasium of St. John's, from which Steinmetz himself had graduated only a year or two previously; and Steinmetz entered the Moser home to tutor George in mathematics.

"I shall never forget the moment when I saw Steinmetz for the first time," recalls Dr. Moser. "Recommended by one of his professors to my father as the best man for the place of tutor, he presented himself one afternoon in our house. There appeared before our eyes a small, insignificant-looking figure . . . clad in the conventional long-tailed coat of black broadcloth."

It was during the tutoring that Steinmetz revealed his youthful convictions on economic and social questions, convictions expressed in the vigorous arguments in which he and Moser engaged. One of these, as described by Dr. Moser, reveals the general trend:

"In speech and counterspeech, we settled the problems of the world easily and grandly, and solved questions puzzling the minds of the most astute statesmen and economists. We were young, and the world was bright. The fire and cocksureness of youth inspired us to limitless flights of fancy, and we were happy.

"Those were the moments when Steinmetz was at his best. I can see him now, his chair swung side-ward from the table, right knee on the seat, bending far over the table's top, firmly anchored on his left elbow, and with his right hand gesticulating and giving emphasis to his phrases, his voice becoming shrill and more shrill the more he immersed himself in his subject.

"When failing breath forced him to cease, I took up the verbal battle, and so it raged, heatedly, fiercely, but always joyously, back and forth, for hours at a time. We agreed almost never; he took his stand invariably with the proletariat, of which I knew naught. He knew something already of practical politics, which to me was a seven times sealed book; in fact, he was so much my superior in most lines, with the exception of music and art, that in these debates I had to be satisfied generally with the loser's end.

"But there was not one instance of bitterness or hurt feelings, not on my part, and I believe not on his; for when the storm was over Steinmetz was again his sedate, smiling self, and only a brighter light in his fine eyes lingered for a while, provided he had not damaged his glasses in the heat of discussion, which happened occasionally.

"Later I came to realize that the conditions under which he had to live were sure to cause violent reactions in a man of his intellectual force. But in those days his impassioned answers to my more or less logical remarks, his wild orations concerning the injustice of the 'ruling classes,' the misery of the workman and small employee, the money evil, the struggle between capital and labor, pleased me, because I had ruffled my friend's feathers.

"This attitude of mind was not surprising, for I did not know anything about these things, brought up, as I had been, always in the midst of my family and friends. But now I know that many things which Steinmetz thundered out in his youth are accepted facts to-day.

"Unfortunately, the young man had come into contact with political elements and had become a Social Democrat, the brand of German Socialist that preached a change in government from the theoretical constitutional to the republican form. The Social Democrats were considered by the German Government then in power as inciters and revolutionists. To be sure, Steinmetz never advocated direct action; he talked only, but used no caution in his utterances, and in this respect he was oblivious to any warning. The time came when he was compelled to leave Breslau between night and morning in order to avoid arrest, and so did he vanish from our home."

CHARLES PROTEUS STEINMETZ

Those were the days in which the Socialism of Charles Proteus Steinmetz sprang forth from the hidden seed, planted—who can say when, or how?

It may be asked, Why did Steinmetz hold such views as those of which Moser obtained so dramatic a glimpse? Principally because he had acquired ideals, ideals which never quite lost their original luster, despite the inevitable disillusionments of life.

First, these were the ideals of the impressionable youth in Germany, with vigorous, unfolding mind gazing about upon a world that seemed to have failed to cultivate its possibilities. He beheld many lives under repression, although others seemed not so unfortunate; while a few sat in the highest places and seemed able frequently to keep down those who were unable to secure a firm foothold on the ladder.

Soon he became altruistic in the sense that he desired to see a fair chance for every one who had apparently failed to get such a chance. It seems to have been a desire to help the under dog, although he realized at the moment that he was a good deal of an under dog himself.

But this ideal, this feeling, became intense with him. It became his creed, his religion. It burned as a consuming flame in his soul, and it burned out from his character whatever germ of self-interest had secured a hold there from the unfortunate spoiling he received when a child. When he first heard the principles of Social Democracy explained in the restaurants and student homes of Breslau, it seemed to Steinmetz at once that here was a practical channel to work through. The goal of Socialism seemed to be his own goal; its purpose, his purpose; its ideals, his ideals.

And thus he became a Socialist, attracted by what seemed the most logical outlet for his philosophy of life. He joined the movement believing, as he always believed, that men and women should give each other the fullest opportunity to make the most of their lives.

To hear this mild-mannered, calmly poised man quietly enunciating his convictions as to the highest welfare of mankind in the realm of social relations, gazing at his hearer with pleasant benign eyes as he talked, and occasionally uttering a deep-throated, appreciative laugh, gave one a feeling that here spoke no prophet of an ill-designing, malignant octopus, to be dreaded by the timid and hated by the dauntless. Rather did it impress one as the voice of a brotherly spirit, yearning to see human destiny ennobled, caring ultimately for nothing so much as the enrichment of life through practical means.

There is a long gap in the career of Dr. Steinmetz as a Socialist, during which he seems to an ordinary observer to be hardly more than an onlooker. It is the gap that stretches from the abrupt termination of his student Socialist affiliation in Breslau, with its over-fervent espousal of the cause, to the year of his placid appearance in Schenectady as a member of the city government, first by appointment of a Socialist mayor and later by election at the polls with the rest of the Socialist city ticket.

Throughout this span of twenty-three years, Steinmetz never for a moment relinquished his principles. These beliefs, to which he clung even though it meant leaving his native land, had naturally not been given up in the interval between Breslau, 1888, and Schenectady, 1911. He had, year after year, faithfully carried his social theories, unchanged in fundamental concept, although modified by the gradual process of time as to their practical aspect. It cannot be said that he brought them into undue prominence, and certainly he did not exploit them in any ordinary sense of the word.

From the general trend of his life during most of this interval, there is every reason to believe that his other pursuits kept his Socialism somewhat subordinated. The absorbing mathematical studies in which he was engaged, with the tremendous volume of intense mental application which they involved, left him neither the time nor the strength for activity in a political organization. Moreover, he was rather keenly impressed, during the first few months of his American residence, with the necessity of earning a living without delay; and, finally, he was not favorably placed for playing any prominent part in the social or economic arena, for he came to this country a stranger, without a friend except the faithful

Asmussen, and quite lacking in an entrée to Socialist circles, had he been disposed to seek anything of the sort.

He found, however, rather early opportunity to acquaint himself with Socialist policies and practices in the United States. As the young draftsman at Eickemeyer's factory, he became friendly with the only other draftsman employed by the establishment, Edward Mueller, at whose home he was later to board for several years. After a few days, Mueller showed more than a passing interest in the young immigrant. He began asking Steinmetz various questions, the purport of which was not particularly apparent. For a few moments Steinmetz answered him unsuspectingly; but, the questions continuing, Steinmetz suddenly perceived that he was evidently being given a cross-examination.

Then the real object of Mueller came out when he began to advocate to Steinmetz, in so many words, the adoption of the cause of Socialism. Steinmetz turned to him, with a quick, whole-hearted smile, and said quietly: "I am a Socialist. I was a Socialist in the University of Breslau, Germany."

Mueller expressed surprised satisfaction; then they shook hands and became fast friends. A few months later, when Steinmetz said he intended to change his lodgings from New York to Yonkers, to be nearer his work, Mueller immediately invited the young man to board at his home. Steinmetz acceded and found the Mueller domestic circle altogether congenial, providing him with a pleasant group of friends as well as a harbor of familiar political principles.

Yet throughout Steinmetz's stay in Yonkers, his brief residence in Lynn, and many years of his life in Schenectady, the world heard virtually nothing of him as a Socialist. It heard considerable about his scientific work but little of his politics.

As his residence in Schenectady lengthened, however, he found ample time for reflection, not only upon the "social issue," as he called it, but upon his personal procedure in relation to it. He emerged as a conservative upholder of the best in Socialism, with a very broad vein of altruism running through his political attitude, thinking always of the greatest good to the greatest number. He advocated the advancement of Socialism, it is true; but by the legitimate, legal methods open to any party.

Moreover, certain characteristics of his political status became more and more apparent from this period until his death. He was an avowed Socialist, affiliated with the Socialist organization. Yet he was also a valued employee of a great capitalistic corporation; and Socialism has always declared itself at war with capitalism.

Unquestionably, Steinmetz was too useful to the General Electric Company, as a master mathematician and far-seeing electrical engineer, for the company to part with his services unless Steinmetz himself precipitated such a parting. And Steinmetz, while he never forgot the inequalities presented in the social strata of the world by the distribution of wealth—which had so burned itself upon his mind as a youth—was invariably consistent, in his own life, in not valuing wealth merely for its own sake. Hence he might not have been tempted by financial considerations to stay in the General Electric organization if he had conceived of his relations with the corporation as a violation of his conscientious political principles.

These circumstances lend all the more significance to his unvarying attitude. For he was a corporation employee more than thirty years; nor did he at any time make a move to free himself from a recognized capitalistic concern.

The explanation is to be found, in part, at least, in his book, "America and the New Epoch." An unbiased examination of that work will reveal at once that Steinmetz believed in corporations. He believed that in economics and sociology the world is passing out of the so-called individualistic era and into what he called the "collectivistic era." He believed that the corporation contains such possibilities of efficiency, and thereby such capacities for usefulness to whole nations—and to the world in general—that it will become more and more indispensable.

Doubtless, he might have felt aggrieved, as a Socialist, toward the individual policies or actions of individual corporation officials. But the corporation as an economic proposition he held to be entirely justifiable. He classified it as the most desirable form of business organization for the present and future needs of the world's history, especially with the approach of an era (which he believed was bound to come sooner or later) when waste must be banished and conservation must be enthroned, if the human race is to survive.

In all this, like the idealist he was, he had in mind the good of the people at large, and gave the corporation—great as it may be, and wealthy as it usually is—a place in the future scheme of things because the corporation, he contended, if it is not misused, cannot fail to accomplish good for the people at large.

Realizing that this was part of his economic creed, and as such was interwoven by Steinmetz into his Socialist principles, it is easy to comprehend why he maintained such a curious status for so many years. He had solid convictions back of his evident contentment in his odd position as a Socialist employee of a capitalistic corporation.

Dr. Steinmetz first actually "went into politics," as the expression goes, in 1911, when Schenectady elected George R. Lunn, the Socialist candidate for mayor. Until that year, Steinmetz had only a nominal affiliation with the Schenectady branch of the American Socialist party. He had kept in the background, quite purposely, it seems; had said little and done less to announce his political principles to his associates.

SOCIALIST OFFICE-HOLDER

But when Lunn was elected mayor on a Socialist ticket, Dr. Steinmetz promptly came forward to offer his services to the new administration in any capacity in which he could be of use. He had never done this before, evidently feeling that he would not have been acceptable to any administration with whose political principles he was not in sympathy. But he was welcomed by Mayor Lunn, who immediately gave Steinmetz a chance to apply some of his practical idealism in a matter-of-fact way.

He enabled Steinmetz to undertake constructive activity for something in which the doctor very much believed, the American public school system, with its "open door" for the children of every American home.

Taking advantage of the action of Dr. Steinmetz in making himself available for service, Mayor Lunn, upon assuming office, immediately named Steinmetz a member of the Board of Education which elected him its president. It was the only situation through which Steinmetz could have attained to this public position, in all probability, for his Socialist party label, it must be admitted, stood in his way. The peculiar combination of circumstances which led Schenectady to elect a Socialist municipal ticket in 1911 was so unusual among American cities that it attracted country-wide attention; the American public at large shied away from Socialist candidates and, in the case of Dr. Steinmetz, continued to shy, as will appear in a subsequent chapter.

CHARLES PROTEUS STEINMETZ

This suggests what must naturally occur to any close student of the career of Steinmetz; namely, that his Socialism, however much sincerity of purpose it may have had back of it, was a barrier to practical accomplishment. Even though Steinmetz made his Socialism the vehicle of idealistic altruism, it suffered from what many persons considered its unfortunate association with the Socialist organization. On the other hand, Steinmetz did an immense amount of good in the world entirely aside from his political activity. He succeeded in accomplishing, as an electrical engineer, the sincere desire of his life to make the world a better place for all mankind. He did much more good as an electrical engineer than he did as a Socialist.

The events which followed Mayor Lunn's appointment of Steinmetz to be a member of the Schenectady Board of Education proved that the electrical genius was an exceptionally fit man for the place. Certain peculiar qualifications for this position soon made their appearance. His analytical mind, his belief in education, his appreciation of the classics as well as the sciences, all enabled Steinmetz to give himself with keen, yet intelligent, zeal toward the solution of the city's school problems.

It may be pointed out that the versatility of Dr. Steinmetz, his Protean characteristic, so to speak, is apparent from his attitude and record in this field of education. It is undoubtedly true that he had made a favorable impression in various educational quarters by several times enunciating broad, sound policies.

His interest in engineering training, typified by his service on the faculty of Union College, was very strong. Yet his views on education were never one-sided. He believed in the indispensable value of a well-grounded elementary schooling and yielded to none in his appreciation of the classics. Several papers on educational topics had borne his signature; and in these, as well as one or two discussions before educational societies, he had taken a firm stand for the promotion of classical subjects in school curricula. His statements had revealed a thoroughly American attitude on the influence of education in national life.

Hence it was by no means a case of inviting an unknown or untried personage to take charge of a city's school system. And there was need for a man with a broad outlook to direct the Schenectady schools at that moment. The city had just concluded a period of rapid expansion, resulting from the establishment in Schenectady of the General Electric Company's main office and largest plant. The city's population had jumped ahead so rapidly that its educational facilities had been outdistanced. Within the twenty years just preceding, while the factories steadily expanded, the population had quadrupled. Yet the school system had not increased in the same ratio.

The entire community needed rather urgently a

man who was something of an educational specialist, a planner, a developer. And such a rôle Dr. Steinmetz succeeded in playing, gaining thereby quite decidedly the confidence of his fellow-citizens.

He found, upon taking up this civic work, a woeful lack of accommodations for the children who were clamoring outside the gates of knowledge. Only one school building had been erected since 1907, and that was in one of the more remote districts, where it was of little aid in relieving the general congestion. There were no less than three thousand boys and girls without seats in the schools of the city.

This condition was the first that Steinmetz undertook to correct as head of the Board of Education. He endeavored to give every child a seat. During his term, three new schools were either wholly built or begun, and two others were enlarged. And when Steinmetz went out of office two years later plans were under way for taking care of every school-child in the city.

He also introduced other innovations, particularly representative of the needs of the schools as he saw them. He worked for and obtained an increase in the number of school playgrounds and playground supervisors, although he was unable to get as much of an appropriation for the latter as he desired. He established a system of medical supervision of the school-children by physicians who served on part time and nurses who served full time. And he worked out a methodical plan for keeping track of epidemics among school-children, in order to make it as easy as possible to stamp them out.

But Steinmetz found the way of a practical idealist in public office in an American municipality was far from smooth. He had a difficult time of it in trying to serve the best interests of the schools. His ideas were not always accepted nor their wisdom conceded. He encountered opposition, political opposition, which manifested itself by giving him trouble in getting the appropriation of funds for school enterprises. The opposition centered in the Board of Estimate and Apportionment, which passed upon the city's expenditures.

This was the secret of the move Steinmetz made, two years later, to win election as president of the Schenectady Common Council. For the president of the Common Council is also a member of the Board of Estimate and Apportionment; hence, if he were elected president of the Common Council, Steinmetz saw that he could secure a vote in the Board of Estimate and Apportionment and work more effectively for the support of his school program.

Steinmetz ran twice for president of the Common Council, each time on a Socialist ticket headed by Lunn. The first time, in 1913, he was defeated when a Fusion ticket overthrew the Socialists under the fervent appeal to the voters to "redeem the city from Socialism."

But two years after that, in 1915, the city having swung back again to a Socialist régime, Lunn was elected a second time on a Socialist ticket, and Steinmetz with him. In that campaign Dr. Steinmetz presented the curious spectacle of a candidate who received the support of the voters even though he did not personally make an active effort. He did virtually nothing positively to win votes. He did not make a single speech. He did not attend a single political mass-meeting. He spent no money. Yet he was elected; although, of course, the Socialist ticket really carried him into office.

In this manner Steinmetz realized his immediate aspiration of becoming president of the Common Council—which he did not care about to any extent, even though it was the only political office to which he was ever elected—and thereby automatically a member of the Board of Estimate and Apportionment—which he did care about a great deal.

"I was not particularly eager to preside over the Common Council," Dr. Steinmetz told his friends, sometime later. "That, to me, was a rather routine, perfunctory position. It did not even carry the voting privilege, except when there was a tie.

"No, what I wanted was to secure a place on the Board of Estimate and Apportionment, to which, as president of the Common Council, I was entitled. Thus I would be able to cast a vote on the appropriation of funds for the schools, as well as for other city departments.

SOCIALIST OFFICE-HOLDER

"To be sure, I was only one. But Mayor Lunn usually supported me in the things I wanted to do for the schools. Moreover, he was able as a rule to influence one or two other votes in the board; so that between us we were successful in voting funds for several propositions and for the more essential enlargements of the school facilities of the city."

And thus it was that he was able to go on with his general program for bettering the school system, after an interruption of two years during which the Socialists were out of office.

That interval had little effect upon the undertakings which Dr. Steinmetz hoped to carry out. The work of building new schools had so far progressed, most of it having been already completed, that these undertakings were not to be halted by a change of administration. The development of medical supervision, the employment of nurses, and the establishment and operation of playgrounds had met with such strong public approval that it was not to be expected the Fusion régime would undertake to reverse these policies, even though Dr. Steinmetz perceived that the Fusionists were not altogether in sympathy with his ideas.

Hence, when Steinmetz returned to office in the city government, he merely went on with his general school policy, with the advantage derived from his double office. He directed the reorganization of the school system, as president of the Board of Education (to which Mayor Lunn reappointed him), and carried on a crusade in the Board of Estimate and Apportionment, by virtue of his office as president of the Common Council, fighting for the financing of the school policies that he advocated.

Among these, the most conspicuous during this period were the establishment of classes for subnormal children; classes for children with tubercular tendencies, to be taught in glass-inclosed rooms on the roofs of the school-houses; a system of ungraded classes for children who could not keep up with their regular classes because of mental slowness, unfamiliarity with English, or recent residence in the city; and classes for anemic children. For all of these Dr. Steinmetz battled in the financial arena, meeting with general success. Indeed, the soundness of his policies is to be judged from the circumstance that none of these innovations were dropped after being adopted.

At the same time, Dr. Steinmetz did not neglect his "perfunctory" duties to the Common Council. He was almost always on hand, to occupy the president's chair when that body was in session. He gave as much attention as was required to the general city problems; but his real devotion was to the schools. And to them went his real energies.

From that time until his death, Dr. Steinmetz was a worker for the schools as president or as a member of the Board of Education. It was his one definite contribution to the civic life of his home city. His service to the cause of elementary education for the community in which he lived was easily recognized as whole-hearted, broad-minded, and constructive. Hence each succeeding administration retained him on the school board; but throughout most of this period Lunn continued as mayor, being elected as a Democrat after his two terms as a Socialist. And Mayor Lunn never failed to reappoint Steinmetz.

Dr. Steinmetz, with his ideal of a society freed from restricted individual opportunity and from what he doubtless conceived to be the mistakes of the old school, always watched with keen interest the development of the Russian Soviet Government. Apparently he had hopes that tangible benefit to the human race in general would come out of the radical experiment upon which the Russian people have embarked, since the overthrow of imperial autocracy.

It was obviously in the glow of such a hope that he wrote a letter to Lenin, in the spring of 1922, in which he offered his technical assistance in the adjustment of any problems of an engineering nature that might arise. He even offered to make a visit to Russia, if the necessity occurred, in order personally to consult with the Soviet leaders. Lenin cordially acknowledged this letter but did not call upon Dr. Steinmetz for any definite advice.

The incident did not add to the prestige of Dr. Steinmetz in the opinion of many persons who fail to sympathize with the Soviet point of view. They were quick to criticize his action, which led them to consider Steinmetz as more of a visionary than a practical idealist, and in any event a man whose political and social views were far from safe.

Yet Dr. Steinmetz quietly continued his interest in Russia and Russian events to the very last. He realized that conditions in that country were abnormal and recognized the existence there of a great deal of suffering among individuals. His reaction to this is indicated in the use he made of a check for two hundred and fifty dollars sent to him by a magazine in payment for an article he had written at the request of the publication. The article in question was a discussion of electricity in American home life. The check he promptly indorsed over to the American Committee on Soviet Relief and sent it to the headquarters of the committee in New York, with the understanding that the money should be applied toward the purchase of tractors for farm development in Russia.

When the Kuzbas Colony was established in Russia, to operate, with American engineering talent, a valuable coal-mining concession obtained from the Soviet Government, Steinmetz, in 1922, was made a member of the American advisory committee. In accepting membership on the committee, he wrote:

"I am very much interested in Kuzbas, and hope much from it. Everybody taking interest in the enterprise knows, or should know, that it is the creation of a better world in which he is taking part, but that he goes out into a field where he must expect hardships and disappointments, where he must organize and create."

That, in a phrase, was the Socialism of Steinmetz —however much it lost standing in the eyes of many by being called Socialism. He hoped for and worked for "a better world." Any idealist who has ever sincerely dreamed of and tried to bring about such a grand achievement, thinking of others, not of himself, is a brother worker with Steinmetz, seeking the thing which Steinmetz sought. And the magnificent consistency about Steinmetz was that, holding these idealistic ambitions, while at the same time having such ability as made him of immense value to his employers in the field of electrical development, he would not seek great pecuniary reward for himself because he believed that "a better world" meant a world in which nobody would demand more money than he actually needed in order to live in comfort.

CHAPTER XVIII

AS the man who tamed the lightning, Dr. Steinmetz was endowed by an extravagant press and a mildly awed public with attributes akin to those supposed to have been possessed by the gods of mythology. Indeed, some of his admirers among the newspapers have habitually alluded to him as the Jove of Schenectady. Enthusiastic cartoonists have pictured him snatching the lightning-bolt from the hand of Jupiter Fulgur and converting it to the use of science. Perhaps the idea actually prevails, in some minds, that he amused himself, when time hung heavy in his laboratory (which it never did), by hurling thunderbolts about the room as easily as boys throw a baseball around the bases.

This wonder-struck attitude of astonishment, with its inevitable exaggeration, grew out of a series of prosaic electrical experiments conducted by Steinmetz and his co-workers, which culminated in the early part of 1922.

It was all in the day's work with Steinmetz. It came up and was carried through in the course of his every-day experimentation at the General Electric Company. He believed it would prove to have a practical side that would be of value to his employers. And such turned out to be the case.

Dr. Steinmetz was a student of lightning for the greater part of his life, chiefly because an investigation of lightning logically followed upon his study of hysteresis and of alternating current calculations. His earlier work made it possible, first, for efficient electrical machines to be designed, and, second, for electric energy in its most general application—alternating current—to be developed into immense electric power systems, whose far-stretching transmission-cables and ramifying distribution-lines carried the services of this tireless giant hither and thither, any-where and everywhere.

But coincident with such tremendous development there appeared new problems, growing out of this very expansion to which Steinmetz contributed so much. He sums up the chief of these new problems in his own simple, pithy words when he says:

"When alternating currents had finally been conquered, and alternating current transmission-lines began to spread all over the country, an old enemy became more and more formidable—lightning. And for many years the great problem upon which depended the further successful development of electrical engineering was that of protection from lightning.

"But before this could be undertaken with reasonable hope of success, we had to know a great deal more about lightning, and the phenomena which are centered in lightning."

This is how Dr. Steinmetz came to study lightning, in relation to its effect upon electric light and power systems. Perceiving that permanent progress was now contingent upon the mastery of this situation, he did not deem it unduly presumptuous to ask questions of the terror of the skies, which has overawed all generations of men through the long sweep of the ages.

Man and nature had already come into inevitable collision when Steinmetz began his investigation of "transient phenomena," under which technical term electrical engineers place all types of electrical discharges that occur only periodically and exhibit erratic behavior. The greater the electrical systems became, the surer they were to encounter interference from the electrical manifestations of a thunderstorm. Ambitious man found that he had assured himself a battle royal with one of the most spectacular elements in existence.

This colossal combat did not, for a while, assume formidable proportions. But that was merely because electric light and power systems had only just begun to spread out their marvelous networks of wires and to produce great machines for sending currents at great voltages over those wires. The provocation, so to speak, was lacking.

As these systems grew larger, however, and the menace from lightning increased accordingly, Stein-

Steinmetz and Einstein, photographed during the latter's visit to America in 1921

metz, as well as other electrical engineers, gave serious consideration to means of safeguarding electrical apparatus from its ravages. He first began making investigations of any extent during the closing years of the nineteenth century and by 1898 was giving a great deal of attention to it.

Steinmetz very soon observed, in his investigation of transient phenomena, that while lightning may have been the criminal which started the trouble in an electrical system, the actual damage and destruction was not done by the lightning but by the normal power of the electric circuit produced by the electric generators. This normal power of the system broke loose and got out of control as a result of the disturbance originally created by the lightning.

With the increasing advance of the electrical transmission systems, from thousands of horse-power to tens of thousands and from tens of thousands to hundreds of thousands, this danger increased also, in proportion.

The lightning-arrester is the particular guardian of electric light and power lines which has been affected by the work of Steinmetz and others who have studied the peculiarities of transient phenomena. This is by no means the same sort of contrivance as Franklin's lightning-rod, designed for the protection of buildings and houses, and still efficient to-day, nearly two hundred years after its invention. The lightning-arrester protects electrical apparatus and electric transmission-lines.

CHARLES PROTEUS STEINMETZ

Probably the lightning-arrester has never, outside the electrical engineering-profession, been given the credit it deserves. If millions of busy people throughout the land understood the way this quick-working device guards them from annoyance by warding off interruptions to the electrical service that is so much a part of their lives, they would look up to it as not the least of the inanimate, ever-vigilant friends of man.

For the lightning-arrester is the sentinel of the transmission system. It is the watch-dog of the electric generating station, the guardian of the power-house. It operates with a swiftness equaled only by lightning itself, and its grand function is to ward off lightning disturbances whenever these outbursts menace the line.

It does this by automatically opening a path by which the vast voltage of a lightning-bolt can jump harmlessly to the ground, instead of coaxing the electric current of the service lines out of all bounds and thereby throwing the whole system into chaos. And after it has shunted off the lightning's energy, by taking advantage of lightning's one inherent weakness—a tendency to follow the line of least resistance—the lightning-arrester prevents the normal voltage of the transmission-line from following after the lightning. It induces the transmission current to stay where it belongs and proceed on its way to serve the needs of man.

This whole little drama of the clash and counter-clash of mysterious forces occurs so instantaneously and is over so swiftly that the people in homes, stores, offices, and factories who are using the electric current that very moment never know that anything out of the ordinary has happened. The lightning-arrester has done its work with silent, unerring efficiency; without being told what to do or when to do it; more dependable than any man could be.

Steinmetz assisted materially in the evolution of the lightning-arrester. He conducted for several years important tests of lightning-arresters, for this was the practical contribution which his lightning-generator brought about. The way this unique apparatus helped in this work will be related further on, but let it be emphasized here that Steinmetz was a decided influence in this, as in other fundamental aspects of electrical development.

In the early days he devoted himself with characteristic scientific fervor to the problems brought on by electrical expansion. His studies of transient phenomena covered, all told, most of the last two decades of his life. Many specific investigations were made, some of which were published as scientific papers.

At length he conducted a systematic study of the general equation of the electric current, which doubtless has little significance to the layman, but was valuable to the profession. It was presented in 1907 before the American Institute of Electrical Engineers. Ten years later, an advanced investigation of the general theory of transients was pursued, and the results issued as a second paper, published in 1917. And in 1919 there was a third paper on the same general subject.

All this work was then collected gradually, systematically rounded out, and published as the book, "Theory and Calculation of Transient Electrical Phenomena and Oscillation." This work of Steinmetz's appeared in its original form in 1909, running through three editions, each somewhat revised and enlarged, the last one issuing in 1919.

But this did not mean that the investigation was completed. In reality it was never completed. Dr. Steinmetz was still busily at work with it when he died; and a large new laboratory-room was in course of construction at the General Electric works at Schenectady, to enable him to prosecute it with vigor.

As his study of lightning progressed, Dr. Steinmetz was able to arrive at some interesting incidental conclusions. Many people had harbored the belief that the enormous electricity of a thunder-cloud could profitably be used for commercial purposes if a method of harnessing it could be discovered. Dr. Steinmetz said: "Not so! The electricity in a cloud that can hurl a thousand lightning-bolts is worth just ten dollars!" Each bolt, he calculated, was equivalent to six hundred million horse-power, for two millionths of a second, which, when divided and subdivided, amounted to a third of a horse-power for an hour; and 333 horse-power for an hour, if multiplied by a thousand—the number of bolts in the thunder-cloud. The market-price for current sufficient to produce 333 horse-power for an hour is ten dollars.

On the other hand, Dr. Steinmetz calculated that there is enough energy in a single flash of lightning, lasting perhaps one millionth of a second, to illuminate a five-room flat every night for over a month, if it could be spread out over a normal period of time. Which is simply another way of saying that a family in the average five-room flat uses in one month, for lighting, electric current equivalent to the electricity in a flash of lightning.

Out of all this logical, orderly study of lightning and its behavior, Dr. Steinmetz came to have a scientist's admiration for the grand scale upon which nature operates in such an instance as one of these great discharges from the sky. He was, in every sense, a student of the thunderbolt. He frankly declared the fascination he experienced in watching a great electrical storm.

And, being as familiar with the habits of lightning as he was, he never felt the superstitious dread of it that troubles many folk. No one ever found him sitting on a feather-bed or hiding in the cellar when the artillery of the heavens was in full play. He would watch a storm of thunder and lightning from his front porch or from his window, if he felt so disposed. But if he had ever been caught in a thunder-storm, out in the country, it is certain he would have walked a long distance and have suffered a severe drenching before he ventured to seek shelter under a lone tree, standing isolated by itself in a field or on a hillside; for that, he well knew, is a real danger-spot in a thunder-storm.

Doubtless this familiarity with lightning-flashes suggesting that he was what might be described as on speaking terms with lightning, was quite enough to set him apart, in the eyes of many, as a person of uncanny temperament; a man who could look a flash of lightning in the face and smile; a man who was on a familiar footing with forces which affright the timid and alarm the sophisticated.

But Dr. Steinmetz would hardly have enjoyed being so regarded. He did not want to be set apart. He disliked to have people think of him as overwhelmingly "different."

This quiet man of science, who never raised his voice when he talked, whose kindly smile had something very winning about it; this philosopher on the welfare of mankind, this friend and playfellow of children—it was not the easiest thing in the world to think of *him* as a "thunderer," or as a "lightning tamer." And this was one's reaction toward him, even after knowing that he really had produced artificial lightning; even after seeing his laboratory lightning-bolt smash the limb of a tree to fragments and hurl them many feet away.

Certain it is that Steinmetz, of all people, did not entertain the slightest contempt of Jove's sky-born artillery. "Lightning," he said, "always has been the most mysterious and most terrifying of all the phenomena of nature."

And then he went on to describe just how it originates; how electrical force is stored up within a thunder-cloud through the amalgamation of countless electrically charged drops of moisture, until the tension becomes greater than the rain-cloud can carry; then the tremendous discharge of electricity known as the lightning-flash, with its attendant deafening clap of thunder.

One day, in the summer of 1920, an unusually accommodating bolt of lightning gave Dr. Steinmetz the "chance of a lifetime" to learn something about its family history, its temperament, and its methods of work. This over-friendly bolt jumped down out of a thunder-cloud and struck Dr. Steinmetz's camp on Viele Creek. The doctor was "not at home" when this occurred; and in his absence the visitor made himself rather free with the property.

A tree overhanging the camp was hit first, the bark being torn from the upper part of the trunk. The lightning then leaped to the camp, where it divided into two forks. One branch passed to the ground through one of the eight-by-ten posts forming the front of the camp. Large splinters were ripped from this post during the passage.

The other fork smashed a window and then hit the camp's local lighting-circuit, which was fed by a storage-battery. Apparently an oscillation was set up in the lighting-circuit, as some of the lamps were destroyed, while others, located between these, were undamaged. The lightning then split into several directions, shattering or tearing splinters from several of the two-by-four supporting posts of the camp, damaging a screen door and a bed. In the bedroom, the lightning jumped from a loose end of wire to a looking-glass, and from there to the ground.

The looking-glass, which has since become somewhat famous, was smashed into many pieces. Some of these were thrown across the room for a distance of more than twenty feet.

As soon as Dr. Steinmetz heard that his riverside lodge had been honored by a discharge of sky-born electricity, he hastened to the camp, where he arrived about twenty-four hours after the visitation. He found that a really remarkable opportunity had been presented for a profitable study of the behavior of lightning.

It was remarkable because it was so entirely isolated. This eliminated any possibility that the evidence might be disturbed before the doctor could make his investigations. He was almost first on the ground and observed everything exactly as the lightning had left it. He could see precisely where the fragments of the mirror had landed. He could observe just how the post of the door was splintered. It was a condition of things parallel to what *Sherlock Holmes* enjoyed—those rare occasions when he was able to reach the scene of a crime before a flock of detectives, police officials, and newspaper men had mixed the trail.

So Dr. Steinmetz set to work, taking photographs and making notes. He was impressed with the explosive nature of the force that had damaged the camp. An enormous outburst of energy, exploding instantaneously, had blown to pieces whatever came in its way.

The most singular feat which Dr. Steinmetz performed, in studying the effects of the lightning, was with the mirror already mentioned. He essayed to collect the numerous shattered fragments and to piece them together, as they had been in the beginning.

A huge task, most people would say! Huge, indeed, but not too colossal to dismay the patience of a scientist when any profitable result is in view.

For hours Dr. Steinmetz occupied himself with this apparent child's puzzle, showing a persistency no child would show. As he proceeded he found that the amalgam on the back of the glass had been partly burned off in such a manner as to exhibit curious markings. These aided him somewhat in fitting the pieces together, a task which was finally completed, allowing Dr. Steinmetz to carry his studies and his deductions much further than would have been possible otherwise.

The genius of the man in working out this remarkable problem, left, as it were, at his door-step, is picturesquely described in the "World Magazine" by William James Smith.

"I offered to help," he relates, "and my efforts must have been laughable to the doctor, as he sat there quizzically watching my ineffectual work in the piecing together of the partly restored glass. Finally, after days of effort, he did piece it together. Then the mirror was carefully placed between two pieces of plate-glass, which were sealed along the edges. In this way it was transported from camp to the laboratory in Schenectady. And now, years after,[1] the result of the solving of the jig-saw puzzle is flashed around the world."

The culmination of Dr. Steinmetz's studies of lightning came in the construction and use of his famous lightning-generator, with which he produced his artificial lightning, to the astonishment of the laity.

It is pertinent to note at this point that Dr. Steinmetz was not alone in his studies of transient phenomena by means of synthetic lightning. In the Pittsfield laboratories of the General Electric Company, F. W. Peek, Jr., had engaged in similar experiments, which were worked out in even more spectacular fashion. There was a miniature village, against which the miniature lightning-flash was hurled, striking buildings which were unprotected by lightning-rods, while those which were fortified with such rods were spared. The voltage discharge of these flashes was higher than Steinmetz's,

[1] In reality it was not more than two years.

attaining two million volts. The purpose of the experiments, however, was different from the purpose of Steinmetz, who was endeavoring to discover exactly what occurs at the moment of a breakdown in electrical apparatus under the strain of an abnormal electrical discharge.

The Steinmetz lightning-generator was in successful operation by the autumn of 1921. Several tests were conducted in the presence of engineering officials of the General Electric Company. Later its spectacular performances were demonstrated to a more non-technical audience. From the beginning it proved useful for the practical purpose which Dr. Steinmetz had in mind, the more accurate testing of lightning-arresters under conditions simulating actual lightning-discharges. This had never before been possible with such a refinement of realism.

Far from demanding entire credit for this unique apparatus, Dr. Steinmetz insisted that his associates should be honored with himself in every reference to it. When he was asked if he would allow a popular description of the lightning-generator to be written, he readily agreed with that stipulation.

"I shall be very glad to help you," he said, "but I am not the only one who has done this. I should like to have the others who worked with me mentioned in your article. It is their work, also."

That was characteristic of Dr. Steinmetz. As a matter of simple truth, he wanted the other engineers brought forward; and they were. One of them was Mr. Hayden, the doctor's adopted son. The other was N. A. Lougee.

As was to be expected, the name of Steinmetz overshadowed those of his co-workers. The public became impressed with the thought that Steinmetz had produced an indoor thunder-storm. One man, instead of three, received the appelation of the "thunderer."

The lightning-generator was not at all extraordinary in appearance. Set up in one wing of the laboratory, it would be quite lost to an outsider amid the assemblage of unfamiliar apparatus standing about the room. A non-technical person would have had to have it pointed out to him.

He would discover that it was absolutely unlike what he had probably imagined. It consisted of several wooden uprights, with crosspieces between them, on which were placed a number of large, oblong pieces of coated glass in several rows, or banks. In a sense, the generator looked like two large racks holding glass slabs. In the original machine there were two of these racks, side by side. Off to their left were mounted several fat glass bulbs of peculiar shape, known as kenotrons, or rectifiers. The rest of the apparatus was merely a network of wires, running this way and that.

There were actually just two hundred of the glass plates in the first of the generators. These were arranged in groups of fifty and in two banks. They are the condensers of the lightning-generator. In the reproduction of nature's theatricals, these represented the thunder-cloud. Like the thunder-cloud, they stored up electric current, bit by bit, noiselessly and unimpressively, just as does the thunder-cloud.

When they were loaded up to their capacity—full to overflowing with electrical energy—they behaved just as the thunder-cloud behaves, on a much smaller scale. They got rid of their electrical cargo by shooting it off, discharging it, through the line of least resistance. And the result was a flash of lightning. It was exactly like a real flash of lightning except in one respect: it was not so powerful.

In the original lightning-generator, the energy of the miniature lightning-discharge was a million horse-power. The energy of a real flash of lightning had been calculated by Dr. Steinmetz at five hundred million horse-power. The voltage of the artificial lightning-flash was one hundred and twenty thousand volts. That of the actual lightning-stroke, according to Dr. Steinmetz's estimate, averages from fifty to a hundred million volts.

Dr. Steinmetz lucidly explained the relation between voltage and energy in a flash of lightning.

"Higher voltages than we use in this generator have been produced," he said, "and have been talked about as 'lightning.' But mere high voltage is not lightning and has no similarity to lightning in its action and effects.

"The characteristic of lightning is high voltage, backed by very large power, lasting for a very short time only, and so giving explosive effects.

"In certain of the high-frequency experiments of Professor Thomson, Tesla, and others, the voltage was very high but with little power back of it; and even in the experiments with a million volts, [1] at Pittsfield, the current was only a fraction of an ampere. Thus the explosive effects characteristic of lightning were entirely absent.

"In our lightning generator we get a discharge of ten thousand amperes at over one hundred thousand volts; that is, a power of over one million horsepower, lasting for a hundred-thousandth of a second. This gives us the explosive, tearing, and shattering effects of real lightning, so that, for instance, a piece of a small tree, exposed to the discharge, is mechanically torn to pieces by the flash. A piece of wire struck by it vanishes in dust.

"The difference is similar to that existing between a pound of dynamite and a pint of gasolene; the pint of gasolene contains more energy and can do more work than the pound of dynamite, but the pint of gasolene gives off its energy only slowly, at a moderate rate of power, while the pound of dynamite gives off its energy explosively, all at once, at an enormous rate of power, and thereby locally tears and destroys."

In all this there was a clear-cut purpose. The

[1] Conducted in September, 1921.

work made possible with the lightning-generator permitted tests of lightning-arresters that were much more accurate than any previous tests. Lightning-arresters constituting, as they do, the watch-dogs of electrical systems, it is of prime importance to know beyond a shadow of a doubt that they will work properly when the time comes. Severe factory tests are imposed upon them to make this certain, yet never before have these tests accurately simulated actual conditions, which occur when a discharge of lightning affects an electrical transmission-line.

But with the Steinmetz lightning-generator actual service tests have been found possible. Lightning-arresters were placed in the path of the artificial lightning and were connected up with the lighting circuit of the building, which they were called upon to protect from the "lightning." In some cases lightning-arresters which seemed satisfactory under the usual tests were found to have unsuspected flaws of operation, and some mysterious failures of lightning-arresters to function properly in actual use were cleared up.

Out of it all has come surer, more reliable protection for electrical transmission-systems and generating-stations.

The lightning-generator had been working for six or eight months before it was exhibited to any one outside the laboratory. After the officials had witnessed its performance, a group of laymen was admitted, and on this occasion photographs were made.

It was about eleven o'clock in the morning when the little party of guests, with the photographers of the company, rode up in the elevator to the top floor of the building and walked down a long corridor, past wire partitions which protected a mighty array of electrical paraphernalia. Mr. Hayden halted the visitors at length in front of the two rack-like frames. They made preparations for the demonstration and for the work of the photographers, answering meanwhile the questions of the spectators.

It was planned first to show how brilliant a flash the lightning-generator would produce; then to exhibit its destructive ability.

The engineers had carefully prepared an easy path, or line of least resistance, for their manufactured lightning to pursue. This consisted of an arrangement of wires which would serve to conduct the bolt. At one point there was a break in the path, and here was placed whatever object was to be the subject of the experiment. By connecting the article with the wire on each side, it was exposed to the full force of the lightning-discharge.

For this first demonstration, a stout section of wire was placed at the break in the conducting wires and connected to the loose ends of the latter. The lightning would pass through it, heating it to a spectacular degree, and showing to the observers—and the camera lens—as a momentary, intense flash.

STEINMETZ AND LIGHTNING

The preparations were at their height and the engineers and laymen were engaged in animated conversation, when a short, quick-moving figure appeared, walking briskly among the laboratory apparatus toward the gathering. It was Dr. Steinmetz, who approached with an alert manner and a cordial glance that seemed to include everybody at once.

"Good morning! Good morning!" he exclaimed; then, without wasting a moment, he quickly drew up a chair and leaned upon its back in the characteristic Steinmetz attitude, his cigar wreathing thin blue smoke in front of his face, while he scrutinized the lightning-generator and the work of his assistants. There he remained, a little apart from the rest of the group, apparently oblivious of the others, but always answering quickly and incisively when his assistants came over to consult him.

He replied to their questions with a quick "Yes, yes," and a vigorous little nod of his head. Only once or twice did he raise an objection. Once it was when the photographers explained what they proposed to do to catch the fleeting passage of his indoor lightning-flash on the photographic plate.

"But how will you get it that way?" he inquired.

The photographers explained more fully. They would do it thus and so. They realized the difficulties; they saw the point that Dr. Steinmetz raised, which, although Dr. Steinmetz was not a photographer, was a very pertinent one. They would do thus and thus to get over the difficulty. That was obviously the only method. Dr. Steinmetz agreed. He repeated his energetic "Yes, yes," and nodded. The preparations proceeded.

Incidentally the photographers who worked with Dr. Steinmetz were impressed by his uncanny ability to figure out their problems in his head. He could always tell them, in a few moments of mental calculation, what size lens to use to get a certain size of picture at a certain distance and at how long an exposure. The photographers worked it out in a matter of minutes, with paper and pencil. Steinmetz could do it in a span of seconds, in his head.

When all was ready for photographing the lightning-generator in operation, the engineers gave the signal, and the current was turned on. The electricity began to flow into the condenser, there to be stored up at a steadily mounting tension.

Almost without warning the "lightning" came. There was a very vivid flash; the wire that had been placed in the path of the lightning stood out for a brief instant, white-hot from end to end; and simultaneously was heard a report like that of a cannon.

Then the indoor thunder-storm was repeated to show its destructive power. A large-sized block of wood was placed in the path of the lightning-flash, and after that a section of the limb of a tree.

The method of placing these objects so that the lightning would strike them seemed prosaic enough. Nails were driven into each end of the block and of the limb. Around the nails were wound the ends of the wires which constituted the path laid out by the engineers for the journey of the lightning-flash. The lightning, they knew, would travel from the lightning-generator along the wire, as the line of least resistance, and upon reaching the gap between the two ends of wire—occupied by the block of wood and afterward by the tree limb—it would leap across that break in its path, splitting the intervening non-conductor in doing so.

That was exactly what occurred. While the spectators stood well back from the range of the bolt's destructiveness, some of them holding their hands over their ears, the synthetic, or manufactured lightning, at the bidding of man, leaped forth with instantaneous flash and bang, and fragments of the block of wood came flying through the air, to fall scarcely three feet away from the staring onlookers. Afterward, as stated, the same thing was done to the tree limb. When the demonstrations were all over, the pieces of the splintered branch were picked up and Dr. Steinmetz held them in his hands, leaning on a high stool, while the camera pictured him thus observing the effects of the miniature lightning that he had succeeded so effectively in bringing into play.

Everybody had something to say about the remarkable imitation of nature they had witnessed. Dr. Steinmetz listened with an appreciative smile and an occasional nod while he puffed at his long cigar. A moment later he disappeared into the laboratory office, where he was soon to be seen leaning upon a desk and pointing out with his pencil various aspects of the particular problem then undergoing examination, a small circle of heads clustered around his own, above the sheet of paper with its mathematical equations and mystical diagrams.

CHAPTER XIX

A UNIQUE ADVENTURE IN POLITICS

HAD Dr. Steinmetz not been a Socialist, he might have been elected to a state office, that of state engineer of the Commonwealth of New York. At least, his nomination for the office on a ticket that included Socialist adherents was the one factor that virtually put him out of the running, the public's reaction toward Socialist candidates being what it is. The nomination of Dr. Steinmetz for this office was on the Socialist and Farmer-Labor state ticket of 1922. It came wholly unsolicited but was accepted by the doctor much in the same frame of mind that he felt when he made himself available for service in the Schenectady city government, with the idea of being of use to the people in any field that presented itself.

At the same time, there were one or two factors in this political venture which would have been quite lacking in the case of any other candidate. The most striking of these, of course, was the engineering and scientific standing of Dr. Steinmetz, who was equipped from a technical point of view far better than any other man whom any party might have brought forward.

CHARLES PROTEUS STEINMETZ

If Dr. Steinmetz had been a candidate for a higher office than that of state engineer and surveyor, his campaign methods would presumably have been different. Yet while seeking election to office he contrived to maintain a dignity which consorted well with his profession and with his eminence in it.

From various motives, he made no active campaign for the office. He went no further than to address one or two meetings in New York, which were arranged by his party managers, and to grant several newspaper interviews in which he revealed in a guarded manner the general trend of activity which might be expected from him in the event of his election. The interviews were printed quite fully in some sections of the press, and in much condensed form by a few of the conservative newspapers.

These views are noteworthy most of all for the emphasis which they place upon the necessity of water-power development within the State. It was in his few campaign utterances of this year that Dr. Steinmetz urged most earnestly that the water-power resources of New York State should be made use of, even down to the smallest streams, in the Adirondack region and in the western section, around the region of the Finger Lakes.

"It is quite conceivable," he told Charles W. Wood, shortly after his nomination, "that the water-power of New York might be made to serve the people." And in another statement, about the same time, he indulged in a bit of speculation as to the ultimate benefit to be obtained, in terms of water-power, from Niagara.

Engineers had estimated before this that there might be a maximum potential volume of energy in Niagara of six million horse-power. Dr. Steinmetz, discussing water-power possibilities after his nomination for state engineer and surveyor, ventured the more extreme opinion that there might be as much as nine million horse-power in the big cataract.

Plainly, he had a mental picture at this juncture of a Niagara given over to power production, with no thought, or at least very little thought, of its preservation as a spectacle of great scenic beauty. He realized that such a proposal brought down upon his head the criticism and downright hostility of a great army of beauty lovers and nature enthusiasts, who differed completely from his own conception of Niagara's purpose, and who would subordinate all else to the esthetic element.

Yet Dr. Steinmetz was himself a lover of nature and all of nature's beauties. And he endeavored to make it plain that he did not yield one whit of his esthetic enjoyment of Niagara's beauty in proposing to encroach upon that beauty for the purpose of supplying the practical needs of a living human race. He attempted to reveal that his paramount thought was for the physical welfare, the physical life and well-being of mankind, so that men might live and develop, as a race, with all the mighty possibilities that an intensely cultivated life might unchain. And to that end, looking into the far future, he was convinced that all water-power sources would eventually be absolutely indispensable. Moreover, he believed the beginning could not be made too soon, realizing that the complete utilization of America's water-power possibilities, or of those of the world at large, will be a matter of many generations in its development.

Interjecting into his views, as crystallized by his impending appeal to the voters of the State, a certain suggestion of purely Socialistic attitude of mind, with its allusion to what should be done for "the people," he nevertheless indicated his long-standing faith in the necessity of harnessing water-power resources, Niagara Falls along with the rest, when he remarked that the complete development of the falls would mean perhaps a revenue of two billion dollars for the State, and, through the State, for the people of the State.

"But," he added quickly, "there is a big difficulty there. To utilize the falls fully would mean to harness the whole stream and to dry up the falls. In the first place, that would mean the opposition of hotel-keepers, because they live on the beauty of the falls.

"But is it worth while to destroy Niagara Falls as a place of scenic beauty in return for one to two billion dollars that can be used for the benefit of the people in a thousand ways? That will be money that can be used to develop parks and public playgrounds and camping sites in all parts of the State for the benefit of the people.

"Is it worth while to dry up Niagara Falls, injure tourist hotel business, and get sufficient money to do things for the people all over the State, that we are unable to do now? That is something to think about."

Before he left the subject, he advanced a suggestion which was in the nature of a compromise. He pointed out that on Sundays and holidays, the falls might be allowed to tumble over the precipice in all their beauty, being diverted for generating electric energy only on week-days.

"It would not necessarily dry up Niagara permanently," he meditated, smoking vigorously as he talked. "We could divert the water for the hydroelectric machinery during the week, and at the close of the week shut down the machinery and the water would again begin to flow over the cliff.

"What a scene it would be, with the water beginning to trickle, slowly at first, then growing until it became the Niagara that we know! Would n't that be a sight even more impressive than we have now?

"Maybe we could do that. Niagara Falls for Sundays and holidays; hydroelectric development the rest of the time."

Truly a prospect to awaken a smile from those who find it difficult to picture Niagara Falls wholly dried up by the hand of man; and then at periodic intervals allowed gradually to resume its grand plunge until finally its full volume was restored—for the time being. The imagination could conceive of eager parties of sight-seers, made up to take in the spectacle of the "return of Niagara," at man's behest. Grand-stand sections would doubtless be erected for the accommodation of thousands who would gather Saturday afternoons to see "the turning on of the falls."

Yet if Steinmetz's previous words are worthy of meditation—and he was preëminently a logical thinker, a prophet of science and economics—that arrangement would be infinitely better than having the falls dried up and never "turned on" again at all, not even for a week-end. That arrangement might well be considered when the day comes, as Steinmetz said it would come, for man to make use of every ounce of energy which Niagara Falls and all the other falls can supply.

So it would seem that Steinmetz really tried to think of the welfare of all the people as far as possible, showing consistency thereby, even when proposing to capitalize Niagara Falls for the production of electric power, at the expense of its scenic attractions.

What Dr. Steinmetz really planned to do, in a practical way, had he been successful at the polls in that election of 1922 in New York State, can be gathered from one simple statement, which was the only definite hint that he let fall; for on the whole he was careful not to commit himself in any published discussion during the campaign. But he did on one occasion announce his intention, if put into office, of making a comprehensive survey of the water-power resources in the State at large, working out a practical policy and plan of procedure, and then "going before the people for indorsement of his program."

He quite frankly declared his hope of accomplishing things, in a general way, which would benefit the people. He spoke with the quiet assumption that folks would understand his sincerity, his honest aspiration to improve conditions affecting the inhabitants of the State.

If many among the professional politicians, many even among the people themselves, noticed an all too-familiar sound to such utterances, Dr. Steinmetz calmly expected that he would none the less, if elected, win the confidence of the people by his deeds, which would speak louder than his words. And he never at any moment labored under an illusion as to the possibility. He realized that he was at a disadvantage in appearing on the ticket which bore his name; but it was obviously a matter of conscience with him—a conviction, which was never shaken.

"I would undertake to make a complete coördination of the various parts of the state service," he said, in explaining more definitely what his program might be, if he won the election. "And one of the big questions would be the use we would make of the water-power we have in the State. I don't know how much water-power we have in the Adirondacks—not much in a single place, but a lot of small water sites—and the problem will be to tie them all up together."

In the midst of his prosaic consideration of the good he thought he could do if elected to a political office which would allow his scientific abilities to function, he unexpectedly drifted into a curious, daring dream, born of idealism and scientific speculation. It was such a dream as he might have often conjured up from a remarkable scientific imagination—to keep to himself. Only in this instance he did not keep it to himself; and so it got into print.

"Naples," he began, with a flickering smile, "has the same latitude as Labrador—yet look at the difference in temperature! It's the warm Gulf Stream that modifies the climate of all of Europe, and makes it much milder than America's.

"Now, there's the Japanese Current, that encircles the Pacific. Bering Strait is narrow and comparatively shallow. If that strait were blown up and widened and deepened, we would be able to divert the whole course of that current north of North America. If that current ran north of our continent, it would melt all the ice and snow of Canada and Alaska, and there would be no more glaciers in Greenland or icebergs in the Atlantic. It would open up a whole half-continent for cultivation and population. It would make all of North America warmer in the winter and milder in the summer. It would double the available habitable globe. It would remake the world."

Then he paused, almost as abruptly as he had begun, and observed that this was merely a dream—as indeed it was; and Charles P. Steinmetz was not given to dreaming. He had once said that it was foolish to dream vain dreams, a thing contrary to the habits of scientists. However, it may possibly be that he allowed himself a different point of view when he was a candidate for office. At all events, this dreamy lapse of an avowed non-dreamer discloses rather vividly the slumbering idealism of the man, which never remained entirely dormant, but was apt to show itself in one form or another throughout the greater part of his life.

It is apparent from the foregoing that Dr. Steinmetz's platform, if such it may be called, contained just one plank—water-power. He stressed nothing else as he did that. In the half-dozen—or less—of public utterances that he made during the autumn of that year water-power was foremost. Frequently it was the sole topic he was willing to discuss.

"Fifty-four million tons of coal are burned yearly in New York State," he said, on one occasion. "Imagine what this means. Filled in coal-cars, it would make a train as long as our two continents, from the northernmost point of Greenland down to Cape Horn at the southernmost point of South America.

"There are fifty thousand industrial establishments in New York State, large and small. Five million horse-power supplies the power of these industries, not counting the steam-locomotives. But of these five million horse-power, three quarters are still produced by steam, from coal, and a little over one quarter comes from our state water-power resources, by electric current. Thus we are only just approaching the electric age, and are still largely in the age of coal—and dirt.

"Yet in California nine tenths of all the electric power comes from water-power, while in our State two thirds of the electric power is still produced by steam, from coal. Yes, you will say, but California has abundant water-power. But so has New York State; far more and far larger water-power sites than most people imagine. I was very much surprised at the large amount of water-power in the State of New York which is still undeveloped and running to waste when I made an investigation of the matter.

One and one third million horse-power have been developed and are being used, saving about thirteen million tons of coal yearly. But over four million are still undeveloped, waiting for the engineer to put them to use in the service of man. There is over a million horse-power in the internal streams, the Adirondacks, the upper Hudson, the Mohawk; and over three million in the boundary streams, the St. Lawrence, the Niagara, and the Delaware."

Charles P. Steinmetz was not elected state engi-

neer and surveyor when the election took place in
November. It was a Democratic year in New York
State, and his Democratic opponent went into office
with the victory of his party.

Yet the strength that Steinmetz displayed at the
polls was a matter of comment all over the State.
For he received a total vote of more than two hundred
and ninety thousand; and in one borough of New
York city, the Bronx, he led the Republican candidate for the office for which he was running. He
was quite decidedly the leading candidate on his own
ticket, measured by the number of votes he received
over any other Socialist and Farmer-Labor nominee.

The election was a defeat for Steinmetz, it is true;
but only a political defeat. It was a peculiarly significant triumph for him as a scientific man, an engineer, a master of things technical, for it made plain
his personal popularity in the minds of a great many
people, revealed the personal prestige which he had
achieved, the knowledge that numerous voters possessed of his career and status in the realm of things
electrical, their admiration for him as a big man in
his profession, and their willingness to intrust a public office to him, no matter what his political creed.

Of course, as has already been observed, he was
preëminently qualified as an engineer. If the voters
had needed only that standard of qualification, if they
had not felt it incumbent upon them to weigh the
political doctrines of the candidates, to ponder the
theories of political parties, they could not have done

more wisely than to elect Steinmetz. On the basis of this purely technical standard of qualification, the satirical comments of a western newspaper editor were both pointed and pertinent. Under the title of "Who Licked Steinmetz?" this editor published the following observation, which is piquant if not absolutely flippant:

"Does any one recall the name of the gentleman who whipped Charles P. Steinmetz for state engineer in New York last month? Probably not many men in that State could tell you right off the bat, not even the fellows who voted the victor into office. Will Rogers, the cowboy-sage confesses his ignorance, but observes that undoubtedly 'he does n't know the difference between a short circuit and a long shot.'

"Whoever he is, the man should step forward and let the world look at him, for any one who has more equipment than Mr. Steinmetz for any engineering job is an important person. However, there is no great possibility that he has or is, for if he had or did his name would be more extensively known. His election is a rather sad commentary on the state of mind of voters who do their political shopping on the strength of party labels. What possible party stamp can be a better guarantee of capacity than the name of Charles P. Steinmetz, what token of vast engineering skill is implied in a party title that exceeds the equipment of Steinmetz?

"There must be one, or the eminent engineer would not have perished politically. Hereafter let it be known, then, that the name of a candidate in Mr. Steinmetz's State is the equivalent of a post-graduate degree in engineering and all other sciences, implying ability beyond that of any man ordinarily regarded as the best engineer in the State.

"And, again, does any one recall the name of the man of this description who licked Steinmetz?"

Before the election, another newspaper editorial utterance, inspired by the candidacy of Dr. Steinmetz, had shown much the same attitude of mind; namely, that here was a golden opportunity to gain the services of a brilliant technical personage, whose politics were really of secondary consideration, especially in view of his own pronouncements on the matter. This editor wrote in the "St. Louis Post-Dispatch" as follows on the point:

"Obviously, an opportunity to enlist a man of Dr. Steinmetz's ability in the public service is a rare one. One of the melancholy facts of American government is that men who have gained great distinction in science or industry refuse all offers of public office. They won't mix in politics. On the other hand, the party leaders shy at the name 'Socialist.' Then, too, they are just a trifle apprehensive about what Dr. Steinmetz might do if elected. They don't know how deep his Socialism goes.

"Dr. Steinmetz's own words may serve to quiet their fears somewhat, if not entirely. He points out very sensibly that he could n't do very much in a revolutionary way. The State Engineer and Surveyor's powers are not the sort that would enable him to set up a soviet government in the State and turn the factories over to the workers. Indeed, his attitude is mainly that of an engineer, as witness: 'What can be accomplished under the conditions that exist depends not at all upon the political belief of the engineer in charge but upon his capacity for engineering. . . . We are all interested in pure water, safe bridges, adequate communications and the elimination of waste. . . . I take it that I have been nominated not as a Socialist, but as an honest engineer. If elected, I should try to serve the people of New York state at least as faithfully as I have served the General Electric Co.' "

Dr. Steinmetz offered no comment on the outcome of the election. He simply accepted the verdict with characteristic tranquillity and went placidly about his business of solving electrical problems and advancing, in every way he could, the welfare of the world in that branch of science in which he was supreme.

CHAPTER XX

HUMAN history, among its many clear-cut revelations, silhouettes the prophet as a man of mental solitude. He who utters genuine predictions, springing from some inner instinct which enables the mind to foresee what will happen in a far-off period by shrewdly gaging the developments of the present, stands most generally quite alone in his discerning vision. He sees what few others see; and because the others do not perceive the picture he has glimpsed, they frequently make light of his prophecy, harking to his words with polite forbearance, but withal smiling tolerantly behind their hand.

Doubtless a good many folk may have had some such attitude of mind when they heard Dr. Steinmetz foretelling that some day we would draw power for our industries from a giant radio-wave, which would belt the globe; or that future years would witness the growing of "energy crops," through which the power latent in the sun's rays could be stored in unlimited quantity, to be applied to the needs of man by being converted into alcohol; or that generations to come would gain their physical nourishment from the scientific breeding of edible bacteria. Those were prophecies, and simply because they were prophecies, everybody felt that there was a chance they might never be fulfilled; for who, they observed, can read the future with absolute precision?

It is, as it happens, a curious circumstance that Dr. Steinmetz himself was wont to decry prophets and prophecies. He once remarked to an interviewer that predictions as a rule represent only so many dreams, and that scientists do not dream, because they see so much practical work just ahead, waiting to be done.

"It seems," he added, in gentle reprimand, "that the only thing the world is interested in is the thing that may be done in future years. The accomplishment of last year fades almost into insignificance. The people are forever demanding predictions. Tell us what you 're going to do to-morrow! And the scientist does n't know that himself!

"If I were to predict that, in five or ten years, we should be able to do, by aid of science, something just now inconceivable, the world would gasp and read avidly my prediction. For me to make that prediction would be foolish, for it would only be a dream; and I do not dream.

"Foolish dreams! What a waste of time and energy to ponder them, when there is so much real work to be done! Little by little, our developments in science make the world a better place in which to

Steinmetz and Marconi chatting on the steps of the General Electric Company's main office building

live, and surely that is what we should do—not waste our time dreaming improbable dreams!"

That does not sound as if Dr. Steinmetz wanted to be looked upon as a prophet. Yet notice his use of the word "dreams." He would not dream, it is true; that is, he would not idly let his imagination run riot, merely to see what wondrous, fantastic picture it would paint, which he could admire, and through which he could attract attention to himself by throwing it before the startled gaze of others. But what he would do would be seriously to ponder the trend of natural development, logically springing from the work of the moment, the probable outcome of which was, to such a mind as his, merely a matter of reasoning.

This style of prophecy has two characteristics. First, it is more likely to be fulfilled and hence deserves far more serious consideration than people are often inclined to give to predictions about the future. And, second, it is not usually quite as spectacular as the far-fetched, unrestrained speculations of those who prophesy more from a sudden surge of enthusiasm than from the precise contemplation of logical probabilities, growing out of known accomplishments already on record.

Furthermore, it not infrequently happens that prophecies like those of Steinmetz are shared in, to a certain extent, by others. Steinmetz was not the only engineer who predicted that radio would be used for power, as well as for communication; nor was he altogether alone in his belief that sun-power would some day be drawn upon in a direct way to keep the wheels turning in man's busy world of industry and commerce.

Thus it would appear that a scientific prophet moves more slowly and more surely than the more sensational prophets, and therefore does not find himself so solitary. Yet Steinmetz more than once seemed so far ahead of his contemporaries as to appear in very truth like the torch-bearer, proceeding in advance of the caravan, throwing his light through the surrounding darkness and waiting for the rest of the world to find its way, by that guiding beam, to the spot where he himself was standing.

It is evident that Steinmetz began peering into the future of electricity from the very moment when he first took up the study of electrical engineering at Breslau University. He was attracted by its vast potentiality even as a young man in his early twenties. The remark he made to another young man of that period, to whom he was acting in the capacity of tutor, seems extremely significant.

"You know," he said to this friend, "electricity is now only at the very dawn, the threshold, to be sure; but hark you, Jork, some day electricity will rule the entire world."

This remained his conviction all his life. The years brought ample justification for his belief, uttered thus so confidently at a period when electricity was far less understood than it is to-day.

Until he began, however, to develop as an electrical engineer, at Eickemeyer's factory, Steinmetz gave expression to no pronounced opinions as to what was to come about in the realm of things electrical. But from those days in Yonkers, he was induced to meditate a great deal upon the future; for the events of the passing moment began to challenge his imagination as never before.

He saw before him—indeed, he himself was partly its cause—the unlocking of many tightly closed doors to electrical development. All men were then speculating as to what would result from the opening of those doors; and Steinmetz foresaw more or less what nearly every one foresaw with regard to the immediate possibilities lying before the profession.

But Steinmetz kept looking still further ahead. During the ten or fifteen years in which he was engaged in making familiar to electrical engineers and electrical students his symbolic method, he was considering what the logical advance of electrical engineering was going to mean for the masses of people. He looked at the question in its economic and social aspects fully as much as he considered its purely scientific side. And he drew upon his training as a physicist and a geologist to scan that mistiest of all horizons, the limitless stretch of whole centuries still unlived. In all this he was surveying the sort of world which he believed would come to pass if electricity developed as he expected it to, and if other events followed in accordance with his gradually formed anticipations.

The earliest results of this continual looking far ahead were to be noticed in his first years in Schenectady. He began to enlarge upon the possibilities of water-power—hydroelectric expansion—which was a subject naturally suggested by the first power-plants at Niagara Falls, then just beginning to operate.

Niagara Falls had already become a name of magic to the electrical engineer who was daring enough to go counter to the lovers of scenic beauty and talk about the water-power possibilities of the mighty cataract. With its latent maximum capability of energy production—amounting to about six million horse-power—it constituted a perennial topic of discussion at meetings of the electrical fraternity. Yet it was so impressive, so startling in a sense, that many engineers of the period were slow in getting beyond it.

Not so, Steinmetz! He realized, like every one else, the immense advantage of utilizing at least a portion of Niagara's hydroelectric capacity. But then he began to scan, in his mind, the map of the entire United States, especially lingering in the mountainous areas, noting how here, there, and everywhere there were miniature Niagaras, each capable of being harnessed, if only the way to do it economically could be discovered.

He visualized the tremendous addition to the nation's power supply which such universal hydroelectric expansion would mean; and he did not fail to consider that the age of steam would not last forever, because all authorities recognized the inevitable exhaustion, some time or other, of the coal deposits.

So he foresaw a day, certain to come, no matter if it was slow in coming, when electricity would be the salvation of a world desperately in need of a substitute for coal. And he also foresaw that electricity itself would be in need of such a substitute, and would find it in the indispensable "white coal" symbolized by every tumbling, foaming mountain stream, the small ones as well as the large ones.

But now there entered into his mind another factor. To place an electric generating-plant wherever a waterfall offered power possibilities, even down to the smallest, would mean an impossible array of trained men to manage and operate the plants; and in most cases these attendants would mean an expense which would far outweigh any economic advantage to be derived from the harnessing of the smaller streams.

How, then, would this situation be met? It was, indeed, a problem! It might have seemed at the moment to be incapable of satisfactory solution. Yet electricity is a marvelous force, capable of the most ingenious applications. And to Steinmetz, who was looking at this whole broad subject as a national necessity, a practical way out at length suggested itself, causing him to become one of the earliest believers in the automatic hydroelectric generating-station.

Years before such automatic power-stations were even designed, much less built, Steinmetz foresaw that they were practicable; moreover, that they would gradually become an essential link in the great electrical systems of the future.

The automatic hydroelectric station is, indeed, a remarkable triumph of electrical engineering, which even at the present time is but just coming into use. Merely a beginning has thus far been made; but the automatic station has ceased to be an experiment. It has finally become a perfected type of small-capacity power-plant.

Its essential characteristic is that it works without attendants. No force of men oversees and manipulates its generators. Whole days at a time there is not a human being present within its walls. In some cases, where the installation is in a wild, mountainous district, the automatic station performs its marvelous functions literally beyond the limits of civilization. It is not, however, left to itself indefinitely; it receives an inspection at periodic intervals, although seldom is anything found to be wrong.

The great economy of such a system is apparent. It permits the small waterfall to be made use of, as well as the large cataract; for the principal cost is in the installing of the equipment. And, as Dr. Steinmetz pointed out, such stations, because of their independence from the indifferent type of attendants which would be available, can function in accordance with the plan of the highly skilled designing engineers who have developed them.

In the operation of these automatic stations, and the application of the electrical energy produced by them, it is necessary, of course, to construct transmission-lines which connect the automatic station with the nearest large station of the system, where the current is distributed according to the demand. Steinmetz understood from the first what this would mean. He perceived the trend that would set in toward what is actually happening to-day, the interconnecting of a number of stations, not only those of the same power company but also those of different power companies. This practice is now summed up in one word, which is potent of the development sure to come within the next few decades. And that word is "super-power."

Super-power had in Steinmetz a definite prophet; for in 1912 he predicted the first step, the concentration of all power supply for a given section of territory into one great system. Fulfilment of this program had begun ten years later, within the lifetime of Steinmetz himself.

About the year 1905, Steinmetz predicted several things that would occur in the slow unfolding of the years in economic conditions in the United States. Among them were the possibility of a scarcity of coal, on account of enormous coal consumption; the exhaustion of the soils of the earth through lack of fertilizers unless nitrate fixation by electricity became commercially universal; the day when electricity will be cheaper than it has ever been before, because of its continual day and night use and its economy of production; and last, but far from least, the indispensable character of the electrical specialist.

It was in an address before the students of the New York Electrical Trade School that he uttered these prophecies. Summarized in his own words, this is what he said:

"When we can use electric power twenty-four hours a day, then we can get it very cheaply. When we can sell the power during off periods, when it is not needed for lighting—during the daytime—doing this without the need of additional machinery, then we can reduce the cost of electric lighting.

"Electric power used during the peak of the load, when the load is highest—during lighting hours—is needed for lighting, and cannot be spared for other purposes. But at other times, when the load is not at the peak, it can be spared, and here is where the electrical engineer and the station operator can do something. He can try to get an even distribution, an even consumption of power, by getting the power used during those hours when there is no need of electric lights.

"If we could find a way to get an even use of power throughout the twenty-four hours, we might be able to supply power at one or two cents a kilowatt-hour. When we have accomplished that, electric power will be much cheaper than anything else, and then the end will come for gas and kerosene. And that time will come.

"But there will also come a time when it will not only be a question of economy, but a question of necessity. At present we still have coal. We use coal whenever we produce electric power by steam-engine, but there will be a time when we will not have any more coal to use, and that is in the not very far future. The anthracite coal will not last very long; and before many generations have passed, even the soft coal will be exhausted. And when the coal has gone, what are you going to do then in the winter-time to keep from freezing? It is a rather serious problem which the next few generations will have to meet.

"When we reach the end of our resources in coal, in the not very distant future, then the only remaining source of power, the only thing which will keep us from freezing, will be water-power, which we will have to utilize electrically. We are gradually extending the use of water-power, but what we have done so far is very little. Consider the case of the Hudson River. We probably use up altogether from its falls a modest volume of water; but we make no attempt to get the enormous power which runs to waste through the spring floods, or in all the creeks and the rivers that feed into the big stream.

"Methods will have to be developed, new ways of collecting the joined powers of all these little streams, creeks and rivers, so as to gather the power together. We will have to do all this when we are at the end of our resources, when our escape from freezing and starvation depends upon our getting the power.

"This is an enormous field for the electrical engineer, and without him there would be hard times coming for the future generations, much harder than we dream of now.

"And there is an enormous, far vaster problem confronting the nations of the earth, which at the present time only electricity seems to be able to solve. In bygone ages all civilization started in the Far East, in the big river valleys of Asia, in countries which are deserts now. These large, dense populations earned their living by tilling the soil.

"That soil does not bring forth any crops now. It is exhausted, has been exhausted for a long time. You cannot get any crops without putting back into the soil in some way whatever you take out in the form of crops. There is no capital any longer in the soil in those regions. And it is getting to be that way in some sections of Europe.

"In America we have been more fortunate. We have had an enormous capital in the soil here. First it was in the Eastern States. But New England is not the farming country that it once was. Its farms are becoming exhausted. There is still the West, with its vast resources, but it is only a question of time when those farms of the West will also come to that pass; and when that time comes, we will not be able to do as we have done in former ages—go west. For when we go west then we shall meet the Pacific Ocean, beyond which are the countless millions of China, whose lands have all reached that state long ago.

"So the last capital is just being used up now, but when that is gone whatever we take out as crops will have to be put back into the soil as fertilizer. For ages, stores of fertilizer were accumulating upon the earth, but we have used them up, until now the last of these, the saltpeter of Chile, will be exhausted in a few decades.

"If all the refuse of our cities could be collected and returned to the farms as fertilizer (although we are not doing this in any manner), it would not replace what we would take out in crops, because there would still be a very large unavoidable loss in the spontaneous self-destruction of nitrogen compounds. Electric power is the only efficient means which at present seems to be able to combine these elements of the air, nitrogen and oxygen, which are necessary as a fertilizer, and which cannot be completely recovered.

"At present we do not use electric power for this purpose to any extent, because we still have our capital of virgin soil, and the cost of electric power is too high. But every year we can see the necessity of producing by electric power a method of restoring the capital to our farms.

"Now that we have so many uses for electric

power, and the supply of the far future will be from water-power, what we will need is a method of completely and successfully collecting all the power which there is in the watercourses of this country. When that is done, there will be no more rapid creeks and rivers, but the streams which furnish electric power will be slow-moving pools, connected with one another by power-stations.

"There will then be no question of saving the beauty of nature when it becomes a question of saving our lives, for that takes precedence over the beauty of nature. We will need electric power then for heating, cooking, and restoring the fertility of our farms.

"Now, all this means that the world needs men who know something of electricity, of the operation and control of electric power. It needs all of us and will need us more every year. The human race will always need electricians. Its very existence will depend upon them."

After great electric power-stations became numerous throughout the length and breadth of the land, after electric power transmission was mastered in almost every phase, and after separate independent power systems began to be interconnected in local, or sectional, super-power systems—all of which were developments foreseen by Steinmetz, although often foreseen in common with other discerning electrical engineers—then the question of increased efficiency arose. And it was during this period that Dr. Steinmetz preached two very definite gospels of electrical efficiency, which embodied distinct elements of prophecy.

One of these was the importance of making electric power generation and transmission an industry by itself; the other was the immense advantage of cobwebbing the entire country with transmission-lines which should be interconnected from coast to coast and from lakes to gulf. This last is the real super-power idea, which several individuals and interests in the electrical profession are now pressing, and to which Steinmetz was lending his utmost support at the time of his death.

Dr. Steinmetz saw very early in the modern period of electrical expansion that the production of a marketable product, such as shoes, machinery, or building materials, ought, in the interests of efficiency, to be kept separate from the production of the power needed to manufacture the product.

"The production of power," said Steinmetz, "is in itself an industrial operation, and in the days of the steam-engine, which operated mills and factories, two industries were really carried on under the same roof—the production of power and the production of shoes, or of breakfast-food, or of furniture.

"This is inefficient and wasteful; the source of our modern high industrial efficiency is subdivision and specialization. A single industrial operation requires all the ability and attention of the administrative, technical, and manufacturing staff, in order to give maximum economy.

"An attempt to operate simultaneously two such different industries as, for instance, the making of breakfast-food and the making of power inevitably leads to lesser efficiency in one of them, and this is the production of power. Thus efficiency requires the separation of power production, just as that of any other raw material, from the industry using the power.

"This is what electrical engineering has done, and herein consists its very great economic value. It has given the world a new industry, the industry of power generation and distribution. Our modern huge electric power-stations, some of them generating more than half a million horse-power, are not lighting-stations any more, but power-generating stations, distributing power for all uses in the territory served by them."

In this, as is clearly apparent, Steinmetz saw that it was a misnomer to call electric generating-stations merely "electric lighting plants," although for a long time most of the output of such stations was used for lighting. To-day the term is passing into oblivion, as Steinmetz foresaw it would.

This trend of thought led into another field, in which Steinmetz predicted some very practical methods. One of these was the utilization of by-product power; another was the location of steam-operated power-plants, wherever a sufficient water-supply for condensation existed, near the coal-mines, thus reducing the haulage of coal.

Particularly earnest was he in advocating the use of by-product power, for this meant waste elimination, and Steinmetz hated waste.

"Energy," he pointed out, "is the raw material of the electrical industry. But energy is also the by-product of many industries. For instance, in the steel plant, great amounts of energy are available in the blast-furnace gases, and to waste this energy and buy power might be uneconomical. Every heating-plant is potentially able to convert from five to ten per cent of the energy of its coal into electric power. Wherever energy is available as a by-product, an electric generator should be installed, converting this energy into electric power, and as such feeding it into the power-lines. Thus, where power is required by an industrial plant, and at the same time energy is produced by it as a by-product, simultaneously power would be taken from the electric lines and fed back into the lines."

Whenever he discussed the theory of overcoming the eventual exhaustion of the great coal deposits by completely developing the nation's water-power resources, Dr. Steinmetz was careful to point out one thing. He declared, with an emphasis which rather amounted to a warning, that even if every last waterfall and every smallest mountain stream were put to work turning generators to produce electricity, all of them together would not create sufficient energy to supply the total demand for power.

This was simply a clear-cut prediction that the nation's water-power resources can never, of themselves, meet the nation's complete power needs, taking into account the normal expansion of future years. Other hitherto unknown sources of energy will have to supplement the hydroelectric energy.

Hinting at the nature of these, in the few years immediately before his death, he went one step further —always marching forward from the known into the unknown—and predicted that biological engineers of the future would develop some new form of plant growth, perhaps a species of bamboo, through which sun-power energy could be stored up to supplement the inadequate power supply available from white coal alone. This, of course, would be the world situation in that spectral period, ever creeping nearer, when the black coal of the earth actually gives out.

The possibility of reinforcing the world's food supply by new and surprising methods also received his attention. On this subject he wrote in 1923, the year of his death:

"The question of feeding future populations opens up a new field for the biological engineer. In the past there has never been any systematic method of producing proteins. The natural protein-producers are the micro-organisms which reproduce at an economically, rapid rate.

STEINMETZ THE PROPHET

"The leaves, year after year, collect the energy of the sunlight, absorb it through the chlorophyl, and produce chemical compounds. We have all around us the collections of solar energy by plant life, but it takes a lifetime for a forest to grow and collect energy.

"Why could not the biological engineers of the future develop new forms of vegetation which would collect the sunlight at a rate hundreds of times more rapid than our present vegetation collection scheme? We have done much in producing new plants, but where each generation takes about a year, in micro-organisms we may produce a generation a day, or several generations a day."

Probably the most discussed utterance of Dr. Steinmetz, in relation to the future, was his declaration that in another hundred years the working day of industry would be only four hours long. Here his practical idealism was coming to the front with confidence; for to achieve any such goal, certain ideal conditions would have to be evolved. There would have to be much greater efficiency in all industrial operations; duplication of effort, wherever found (even the duplication that exists in trade competition) would have to go; and in its place intelligent coöperation would have to prevail.

And Steinmetz never had in mind a day made up of four hours of productive work and twelve hours of lazy idleness. He pictured something far more stimulating than that. He looked forward to an era of such quickened mental development, such advanced and universal education, that people would spend much of their twelve hours of leisure in cultivating their higher natures, in acquiring uplifting knowledge, in seeking soul-awakening recreation, as well as the upbuilding of physical health.

Few people took this prediction of Steinmetz seriously. Yet there are those who dare to believe that Steinmetz was simply ahead of his day, and that it will be interesting to know, four or five generations hence, to what extent the business and industrial leaders of that period have come around to his way of thinking.

"Work," remarked Steinmetz, incidentally, "is a curse. The chief aim of society should be to abolish work."

"And what he meant by 'work,'" writes C. W. Wood, "was drudgery, of course; but the 'drudgery' he had in mind is the 'work' of thousands. He hopes for a day when every man will be able to work creatively."

Finally Dr. Steinmetz turned his attention to the new and magical field of radio. Although not a contributor to radio development, he saw in it possibilities as fascinating as any that existed in the application of electrical energy transmitted by wires. He saw in it a possible new method of transmitting power, a method that might have extraordinary advantages over the old ways. This, it is true, was not a question of power creation but one of power transportation.

One of his last utterances relating to future years was on this subject. Several times in the last year or two of his life he hinted at his belief that in some not far distant period power for the world's needs will be obtained from a giant form of radio-wave. In this realm of speculation he had ceased to engage in mere questioning and had asserted an opinion that clearly inclined toward positive prophecy.

He suggested that the world will get its future driving-power from great power-sheets, belting the globe on radio-waves that will be measured in millions of kilowatts and that will be sent forth from one tremendous central power-station. He pointed out, however, that before such a colossal situation could ever be attained, science must first discover the super-electromagnetic wave that will make such gigantic power transmission possible.

CHAPTER XXI

RAILROAD electrification was one of the transformations of the rapidly developing electrical age in which Dr. Steinmetz displayed a tremendous interest. Repeatedly he had predicted the inevitable approach of the electrified railway system, which would supersede steam railroads. He looked upon it as one of the next great forward steps toward utilizing the full possibilities of electrical energy.

In 1920 he mentioned electrification as the step which would do more than any other one thing to overcome the vast operating problems of American railroads. In 1922 he said: "I believe the railroads of the United States will soon be operated by electricity. The change from steam may come any day. One of the big systems will start, and the rest will be compelled to follow."

Then he added a truism that is known to all electrical engineers and accepted in theory by nearly every one who has closely studied the subject: "Steam cannot compete with electricity. It costs more and does less. A steam-engine must slacken speed on up-grades. Electric locomotives, with unlimited power behind them, can go at top speed all the while."

Steinmetz never played a prominent part in railroad electrification work. He was called in consultation on problems connected with such developments, as he was whenever a difficult problem presented itself. But it was not a field to which he made any direct contributions, as he did to the field of street-lighting, or to the basic question of economical current distribution.

Being a student of every phase of electrical engineering, however, he studied the application of electricity to railroading in a thorough manner; and the inherent wastefulness of the steam-engine centered his attention immediately. For Steinmetz was a foe of economic waste, wherever it showed its head. And he considered railroad electrification as merely one aspect of an era soon to come when wasteful methods will vanish before the advance of electricity, the economizer and efficiency builder.

He pointed out, also, that so far as electrical science was concerned, America's railroads might all be electrified to-day. "Electricity," he said, "has been ready. The railway managers were the ones who were not prepared to move. Their finances were not in a favorable condition. They could not borrow the money to make the change. Moreover, they were not sure that they wanted to adopt electricity as their motive-power."

This attitude appeared to a mind like that of Steinmetz as rather over-conservative. To his view, judging the matter simply as a question of ultimate efficiency, and convinced that electric energy would provide such ultimate efficiency, this attitude seemed like living in a rut.

Steinmetz had gathered a few towering statistics on this question of railroad electrification and what it would mean. He presented them, as part of the case in favor of railroad electrification, in "Hearst's International Magazine" for May, 1923.

"Railroads," he said, "burn in their locomotives about 160 million tons of the coal they carry. If the railroads were electrified and coal for manufacturing purposes were burned at the mines, it would be approximately equivalent to doubling the freight-carrying capacity of our railroads for other kinds of freight. About half of the freight now carried is coal.

"It should mean a good deal to this country to get rid of the steam locomotive and the coal train. They are both wasters. Whatever is wasted anywhere is a burden upon the country. It is lost motion to devote approximately half of the freight-carrying capacity of our railroads to the transportation of coal that should be burned where it is mined whenever conditions make this possible, and converted into electricity."

These views were only a part of the story, as frequently related by Dr. Steinmetz in vigorous, vivid phrases. In this instance he was picturing electricity as the giant of the ages, capable and ready to do anything under the sun that might be required of it.

"What will electricity do for us yet?" he asked. And then, answering this pertinent question: "It will do whatever energy can do for us. Nobody, to-day or ever, can fix the limits to which the use of electricity may go. We can say only that it will go as far as human need for energy goes. Electricity is energy, and energy is the basis of civilization.

"We call this the age of electricity, but it is n't. The age of electricity has n't begun. All that we have yet done is but preparatory to the ushering in of the electrical age. When the age of electricity comes—as it will—electricity will do for everybody all that it can do for everybody. It will do all this in addition to doing a multitude of things of which we have not yet dreamed.

"I came to America in 1889. It seems a long way back to think where the development of electricity was at that time. It seems a long way ahead to think where it will yet be. For the age of electricity is yet to come. And it will be a great age."

A day when the familiar steam-locomotive will no longer puff and chug its way from town to town, and thunder over the great highways of steel, half hidden in wind-blown streamers of white vapor, is not easy to imagine. Steinmetz could imagine it because he was accustomed to "viewing the whole physical world as a unit"—exactly what he once declared every student of electrical engineering must learn how to do.

As previously observed, he realized that railway electrification was but one side of a future all-electrical world. But to the people of this generation, who listened to his predictions with sometimes more indifference than the situation justified, this seemed to be the most astounding statement that he uttered, the most astonishing single metamorphosis which he affirmed electricity might be expected to achieve.

It was a prediction that Steinmetz not only never hesitated to pronounce, but one which he even applied, upon occasion, to localized districts, as if welcoming an opportunity to impress people by describing how it would affect their home environment. Thus, he told an interviewer who hailed from New England that in the New England of the future "every steam-engine will be scrapped."

"The time will come, and in a very few years," he reiterated, "when electricity will completely take the place of coal. Power will flow through a network of wires to every engine and factory in New England, and that power will be generated from the flowing water of the territory."

Steinmetz had already announced that he believed New York State possessed 4,200,000 horse-power of potential water-power energy. He tentatively declared that New England probably had 5,000,000 horse-power in its waterfalls and streams. What 5,000,000 horse-power signified was disclosed in his explanation:

"Every horse-power means ten tons of coal, more or less. On railroads where steam-locomotives are used, more than ten tons of coal are used to generate one horse-power. This is due to the inefficient system of our steam railroads, where heavy locomotives have to haul their own fuel.

"You will see that with an estimate of five million horse-power available in New England, this would mean a saving of fifty million tons of coal a year— and that would mean shutting down every steam-locomotive in the territory. This is not a dream, but it is a fact that must be faced. This development will come.

"A saving of fifty million tons of coal in New England is hard to realize. Do you know how much coal that is? If it was loaded on coal-cars, placed end to end as they are in a train, it would reach from Boston to Cape Horn, in South America, and there would still be several million tons left over."

To enable his visitor to visualize a future country with electricity as the universal servant of man—noting that thus far it is by no means universal in its service—Dr. Steinmetz then said:

"If you will picture to yourself a map of the United States showing all of the railway systems, you will get a picture of what must happen to New England. The network of railroads shown on the map can be likened to transmission-lines between the hydroelectric plants which will dot New England.

"These lines will be double and treble lines, passing from city to factory to railroad to hydroelectric plant. It will all be one vast network, some points taking out power and some points putting in power. That would mean the scrapping of every steam-engine in New England."

Turning again to the railroads, he pointed out, as he usually did when discussing all these possibilities, that the operation of an electric railroad, which used hydroelectric power, would inevitably be cheaper than the operation of a steam railroad, because the primary source of the power—white coal—is furnished by nature and costs nothing after once put to work. The initial cost of the electrification would be high, but after that the expense would be insignificant by comparison. It would mean not only saving half the coal now required, but also an increase of as much as twenty-five per cent in efficiency, through the higher speed and better control of the electric trains.

These observations and recorded comments of Dr. Steinmetz upon electricity as applied to railroad transportation are perfectly typical of the way he talked upon this absorbing subject. Perhaps it may seem as if he had more to say about railroad electrification than about the electrification of almost any other one industry. But this may be accounted for by the probability that the steam-locomotive impressed him as the most common and most unmistakable example of economic waste in modern society. He would not have discounted the value of the steam-engine to the world of James Watt, which previously had never known this tremendously significant piece of machinery. His only indictment of the steam-engine was this, that in the presence of the electric generator and the electric motor the steam-engine represented by comparison an inexcusable inefficiency. In its day he recognized it as a great developer of civilization; but its day, said Steinmetz, has begun to pass.

"Steinmetz," said a writer who summarized his career shortly after his death, "foresaw an era of electricity. To him a ton of coal was not merely a costly and scarce commodity for heating the house or cooking dinner. It was so much potential energy, imprisoned by nature, waiting for man, its master, to release and put it economically to its proper work.

"In a ton of coal, as it is taken from the mines, loaded on the cars, brought to our doors, and burned in our furnaces, he saw appalling inefficiency and waste. He saw in all the processes embraced in mining, marketing, and burning that coal a consumption of time and energy which could be put to a myriad better uses. His dream was of a time when the coal will be taken from the ground expertly, cheaply, mechanically, and burned, as far as possible, with far greater kinetic energy than at present, at the place where it is mined.

CHARLES PROTEUS STEINMETZ

"Steinmetz's bold imagination went beyond the era of coal. . . . He looked to 'white coal' to supply the future's needs. Water-power is inexhaustible. It is the refuge of a future generation, which would revert to savagery without means of heat, light, and motion. He wanted to bring the device of that distant to-morrow up to the period of to-day."

In another realm of modern transportation Dr. Steinmetz came forward during the last decade of his useful life with an equally positive view as to the ability of applied electricity to replace inefficiency with efficiency. He began making a study, sometime between 1912 and 1918, of the automobile truck.

Few people had, at that time, much notion of electric trucks. Gasolene trucks were accepted as the natural development, because everybody knew how universal the gasolene motor-car had become. But Steinmetz realized that with the passenger automobile very little account is taken of the cost of operation. There is no element of profit and loss; it is a pleasure-vehicle in most cases, and the cost of its upkeep does not go down on an office ledger, to be tabulated at the end of each year either as an asset or a liability, as a promoter of a gainful business or a drawback to that business.

Believing in electricity for every walk of life and every department of human activity, he proposed to apply electricity to trucks, partly as a matter of course, as an experiment to see if it compared favorably as to operating costs. He finally came to the conclusion that for short hauls the electric truck would prove superior to the gasolene truck, and he began to express this view.

Eventually he went further, much further. He finally was induced to enter upon the designing of a new electric truck of his own invention. He believed that a remarkably efficient electric truck could be produced, and the possibility fired his inventive imagination, so that he finally set to work in his laboratory upon this idea.

From this activity there came forth—on paper— the Steinmetz electric truck. It made more intensive use of electricity than any other truck that has previously appeared, and hence carried a larger number of storage-batteries.

Steinmetz was eager to see his truck put to actual use, or at least given a practical trial. It was soon given such a trial, proving to be a vehicle of unquestionable power. It seemed especially to be a good hill-climber. The performance was on the whole very satisfactory to Dr. Steinmetz.

A few years before his death, in 1921 or 1922, a group of business promoters undertook to manufacture and market the Steinmetz truck. Dr. Steinmetz was given an office in a corporation organized for this purpose, although his post was honorary only. But the business organization was by no means as efficient as the trucks themselves. There was an evident lack of good management. The venture, to the misfortune but not the discredit of Dr. Steinmetz, did not prove an enduring success. And he was wholly without business ability himself; his genius lay in quite a different direction. Consequently the Steinmetz electric truck was not produced in any quantity nor did it become popular among business concerns up to the time of its inventor's death.

Dr. Steinmetz remained as much convinced of the superiority of the electric truck over the gasolene truck, however, as he was of the superiority of electricity over steam for railroading. His views on short hauls by electric trucks as compared with short hauls by gasolene trucks were summarized by him, at the time he brought out his own electric truck, in these words:

"Reductions in the cost of delivering merchandise would be distinctly a service to the public, for the merchant who effects such savings is usually able and eager to pass them on to his customers. One of the greatest evils of our economic system to-day is the high cost of distribution, and I am convinced that a greater use of the electric truck will mean a considerable decrease in the cost of distribution, especially of food commodities.

"It has always seemed a surprising thing to me that electricity—the greatest driving-power known —has not yet been given its full position in the transportation world. I do not think I exaggerate when I say that the use of electric motor-trucks for short hauls will mean a saving of millions of dollars to the people of the country."

How the views of Dr. Steinmetz impressed others, as reflected in newspaper comment, is interesting as showing the reaction of the lay mind to his economic and electrical predictions. A commentator, who was plainly impressed by the Steinmetz version of an electric world to come, thus sketched this world of the far to-morrow as he conceived of it, writing about a year before the death of Dr. Steinmetz, and confining himself for the most part to statements that Steinmetz himself had made:

"He [Steinmetz] says that our present use of coal is absurd, and that it is an economic crime to burn coal for mere heating without first taking out of it its mechanical and electrical energy. The time is not far distant when much of the coal will be burned at the mouth of the mine, and instead of transporting the heavy coal we will simply transport the energy in the shape of electricity.

"Very probably our children's children will see the last of railway locomotives carrying their own coal. All that power will be brought to them by electric wiring.

"He reminds us that we have also hardly begun yet to use our hydroelectric power. Indeed, an entire transformation of our material life is very close at hand.

" 'When everything is electrified,' he says, 'and central stations are automatically coaled, smoke can be eliminated. No dirt, dust, or smoke will be permitted by the Government. There will be no fires of any kind within the city limits. Hence there will be no conflagrations, and the air will not be full of gases.'

"The streets will be clean. There will be no animals used for traction, and hence no street dirt.

"The atmosphere will be clear, and you can always see the sky. With pure air, the yards in the city can be improved and parks beautified.

"In a word, all of the power which we now get by dirty and wasteful processes will be manufactured at waterfalls, coal-mines and oil- and gas-wells; and the three great essentials, heat, light, and power, for human activity, will be attended to by the genie which Ben Franklin invited from the skies."

CHAPTER XXII

HIS LIFE IN OFFICE AND CAMP

FOR the last fourteen years of his life, Dr. Steinmetz stood first on the list of consulting engineers connected with the General Electric Company. They were years of conspicuous study in all the paramount electrical investigations of the period; years during which his advice was sought repeatedly by engineers who were in the vanguard of the advance which electrical science steadily carried on. Steinmetz himself did not carry the torch of progress in his own hands quite as much as he did in earlier days. He was now preëminently just what he was called, a consulting engineer.

He was still engaged in electrical research work, however. This was a lifetime interest, something he never could have relinquished. All that he did in his later years to investigate the behavior of lightning, as it affects large electrical systems, was simply an item in the program of scientific investigations which he followed in the last decade of his career. He was striving, at the time of his death, to discover more than had ever yet been learned as to the operation of lightning-arresters, concerning which he had already revealed enough fully to justify his famous lightning-generator and its spectacular performances.

The definite object of his last line of research was to find out just what happens at the moment of an electrical breakdown, when a transient discharge—lightning or an induced surge of electric energy in a circuit—occurs. He had intimated to some of his associates that he hoped for bigger results from this study than he had obtained from any of his earlier investigations.

It seems difficult to imagine anything he could have done which would have been more valuable than his law of hysteresis and his theory of alternating current calculation, or his invention of a system of alternating current distribution; yet he seems to have discerned with his far-seeing scientific eye even greater accomplishments, which, unfortunately, he was not to be allowed to consummate.

He was placidly happy in his work throughout these years of mellow maturity. He had every possible facility; and he followed his own peculiar bent in conducting the work of his office as much as the work of his laboratory.

His office was in the headquarters building of the General Electric Company at Schenectady. There he had a large desk, with a huge, leather-seated chair, which, incidentally, no one ever remembered having seen him occupy. His favorite posture while at work here was to stand with feet slightly spread apart and lean with both elbows on one corner of the desk, calmly reading over his correspondence (his conception of real work) or conversing with whoever had come in to confer with him—and invariably smoking a cigar. He had his own private secretary, who did all his typewriting and manuscript work.

Dr. Steinmetz orally dictated hardly a dozen letters a year. He used instead his own original shorthand system, by means of which he wrote out in the form of notes what he wanted to say, and then handed the notes to his secretary for transcribing.

This was one of his several minor hobbies, the invention of his own shorthand alphabet. Few men less versatile would have even conceived of such an innovation; and the young women who served as his secretaries at one time or another were likely to stand in awe of the doctor's peculiar office methods as much as of the man himself. His secretary was obliged to study and become an adept at reading the doctor's shorthand, in addition to the commercial system she had already learned in business school.

Dr. Steinmetz developed his shorthand gradually over a period of years, beginning with the days when he took notes at his lectures at the University of Breslau. Its foundation is the Swedish Arends system, combined with the best of several others and with modifications of his own devising. The underlying principle is entirely phonetic. He wrote the word "height" in three letters, "h-i-t," with a long "i."

"Once you learn this system you will never forget

it," Dr. Steinmetz used to say. "It is like swimming in that respect. With other systems a person who does not use it for a year or more is likely to forget the greater part of it. But my shorthand is simple, and as readable as longhand. I can read to-day notes that I took forty years ago, and read them as easily as those I took forty minutes previously.

"I took all my notes in college by this system, or by its underlying principle. It was so much faster than longhand. And I can still read those same notes, all of which I have in bound volumes. I don't know of any other way I could have done this.

"My experience has taught me that shorthand is of great advantage, especially to students. The work becomes simple with shorthand, since the student's attention is not detracted from the speaker by his writing, as is often the case when one attempts to make notations in longhand."

Dr. Steinmetz used to contend that his system of shorthand enabled him to write as fast as he could think. It was undoubtedly a great time-saver for him; one page of his shorthand notes was equivalent to three pages of double-spaced typewritten manuscript.

His invariable method of answering his mail was to make shorthand notes at the head of the letter, or on the margin, indicating exactly what he wanted to say, and then handing the letter to his secretary, who wrote the letter and returned it to him for signature. Sometimes he would give her whole manuscripts in shorthand and allow her three or four days or a week to do the transcribing.

Once in a while these manuscripts would consist of page after page of mathematical equations, frequently of the most intricate nature. Most stenographers would have found it sufficiently arduous to master the Steinmetz shorthand without the added nightmare of the Steinmetz mathematics. In fact, one of his later secretaries describes the experience of "breaking in" under Dr. Steinmetz, and the mental sensations upon first sight of one of his manuscripts dealing with a mathematical subject.

"When I was assigned to take up work in the consulting engineering department," she says, "it was without the most remote idea that I was to work for Dr. Steinmetz directly. My confusion at seeing his name-plate on the office can hardly be imagined.

"A long manuscript, the revision of 'Transient Phenomena,' was the first piece of work I did in Dr. Steinmetz's office. This book comprises numerous mathematical equations, which in the manuscript were filled in with pencil. I had never had occasion before this to write equations, and had not the slightest idea how it was to be done. But by 'juggling' the carriage of the typewriter and using a great deal of care and also time, I was able to hand in sheets satisfactory to the doctor.

"My time in Dr. Steinmetz's office was extended, because of the continued absence of his secretary, but I did not see the doctor for several weeks. I had never seen him before this, and what was told me of his greatness filled me with consternation.

"This feeling, however, turned out to be quite unwarranted, for when he did come the atmosphere instantly cleared; and ever after that it was with very different anticipations that I awaited his arrival at the office. He was always ready with the utmost consideration and willingness to assist me whenever I met with difficulties in my work.

"From the first morning that Dr. Steinmetz entered the office his greeting was always the same, and he never missed it. He would say: 'Good morning! What's new?' with a cheerfulness that made everything seem bright the rest of the day, no matter how gloomy it was outside nor how disconcerting the work inside."

The inventions of Dr. Steinmetz, all of them in the technical field of electrical methods, were scattered through the greater part of his career. He worked on some of his electrical inventions as early as his days with Eickemeyer in Yonkers; and he had some undeveloped ideas sketched out almost up to the day of his death.

A total of 195 patents stood in his name. Of these, more than forty were related in one way or another to the broad subject of systems of alternating current distribution, the notably useful invention which he himself always declared was the most significant of all his patents.

LIFE IN OFFICE AND CAMP

The particularly valuable inventions on the list were those relating to the monocylic system, as it is technically termed, these constituting the nucleus of his inventions relating to distribution systems; the tandem connection of alternating current motors; the application of the tandem or cascade connection to frequency conversion; the induction regulator; phase transformation; compensated alternating current motors; the flaming arc electrode containing titanium carbide; a seal for fused quartz lamps; and several others of less importance, in addition to the patents on the magnetite arc-lamps.

These inventions did not contribute immediately to his fame, either in or out of the profession, so much as did his mathematical achievements, yet with him they were all placed under the broad classification of electrical engineering. And often the seemingly divergent activities which he carried on in his laboratory at the General Electric works or at his residence laboratory were the forerunner of one of his purely inventive contributions to electrical science.

More than once he was found by his visitors immersed in laboratory undertakings. Charles W. Wood presents a typical picture of Dr. Steinmetz at such a moment, in an article entitled "The Romance of an Engineering Genius." [1] He writes:

"Dr. Steinmetz had worked himself to the very top of the electrical engineering profession when I first came in contact with him. It was not as a newspaper

[1] "National Pictorial Monthly."

man that I first met him. I was brought to his laboratory by a machinist who wanted to ask him a question about mechanical principles which was too deep for a mere newspaper man to understand.

"I ventured a few questions about the schools in Schenectady (he was then president of the Board of Education under the first Lunn administration) and was surprised at the simple way he answered me. I discovered that I had been afraid of him; I was afraid of his knowledge, afraid of the big head which had solved a thousand mysteries and which must look with contempt, I thought, at the rank and file of cheerful idiots about him. I said something, in fact, concerning a hope that he would not consider my questions foolish.

" 'There are no foolish questions,' he said, with an engaging grin. 'No man really becomes a fool until he stops asking questions. The science of education is the science of helping people find out what they want to know.'

"Which is one of the reasons why I call Charles P. Steinmetz the greatest man I ever met. He was so much more than a specialist, although he was so much of a specialist that he wrote text-books which only the most highly trained specialists in his particular line could read. Yet he denied that he was a 'wizard,' and had no desire to play the rôle. What he had learned about electricity, he pointed out, was due to his being curious about it. He simply asked

Steinmetz and Thomas A. Edison, standing in front of Steinmetz's famous lightning generator during visit of Mr. Edison
to the General Electric works at Schenectady in 1922

questions, and one question led to another, and there he was."

Those who were associated with Dr. Steinmetz in engineering work at the General Electric Company always knew they could count on securing his assistance when it was needed, for he was never too busy to engage in an engineering consultation. Nor did he ever scorn to aid another engineer in solving a mathematical difficulty. Once or twice he could not resist the temptation of showing another engineer, uninvited, how to overcome a baffling problem, even though he might give the appearance of exhibiting superior knowledge unasked. There is one somewhat famous incident of this sort, which, although it has been recounted a number of times, yet shows the sincere desire of Steinmetz to be of use, in small matters and in large, and does not indicate any motive of overweening superiority. The doctor was by chance passing the desk of a designing engineer who was puzzled by a mathematical problem which he had laid out before him. Glancing almost casually at the equation, Dr. Steinmetz, exhibiting his characteristic marvelous mental quickness, grasped the whole situation in an instant. Then he said to the engineer: "If you will now put your x in place of your y, and your y in place of your x, you will have the answer." Then he went on to explain the process in detail.

On another occasion he was making an investigation of some new engineering applications, and was not only away from his office but out of reach, at the moment, of any engineering or mathematical records which would give him the data employed by engineers in their calculations. He found himself in need of the table of logarithms, but without any possible way of securing access to a book that would give it. Whereupon this super-mathematician proceeded to compute the table of logarithms from memory, and to apply it to the problem in hand, all without aid of paper and pencil but simply by visualizing it all in his mind. Little wonder that those who witnessed such a performance were inclined to bestow upon Steinmetz the awe-inspired title of "wizard," a title which he always disliked and never encouraged.

He frequently appeared at various conventions of electrical engineers, illuminating engineers, or scientific bodies. At these assemblages he encountered some of the other great minds of which he was a contemporary. He had met Edison a number of times before the latter's noteworthy visit to the General Electric works at Schenectady in the autumn of 1922. He had met Marconi, Tesla, and others, and was a delighted chum for several days of Elbert Hubbard, whom he also met on another occasion, at Association Island, on Lake Ontario. This was in September, 1913, during a council of electrical leaders to discuss the question of coöperation in the electrical profession. The impression Elbert Hubbard gained of Steinmetz on this and other occasions is vividly portrayed in the following appreciation of some of these electrical pioneers, published in Elbert Hubbard's

"Fra" for December, 1913, and including not only Dr. Steinmetz, but also Samuel Insull and Henry L. Doherty:

"Steinmetz, next to Edison, is our great modern mechanical prophet. Steinmetz seems possessed of faculties beyond the average man. He has an intuitional sense that is almost uncanny.

"His 'boys' may work on an electrical problem for a year or more and fail to make it tangible. Steinmetz will then sit down and look at the machine for about five minutes, light a cigar, blow a cloud of smoke through it, and behold, the thing starts and chaos becomes cosmos!

"The subtlety and keenness of the man's power, with his ability to talk lucidly, logically, simply and sanely, mark him as one of the world-makers.

"When Dr. Eliot, then president of Harvard University, conferred the degree of Master of Arts upon Steinmetz, he did it with the words: 'I confer this degree upon you as the foremost electrical engineer of the United States, and, therefore, of the world.'

"If in some respects he has gone beyond Edison, the fact must not be forgotten that he has built on the master. Edison had not only to discover the principles of electricity, but he had to manufacture the machines to control the current.

"Well did Steinmetz say that in untamed Nature electricity is the most useless thing you can mention. Without the genius of man it is purely destructive in its nature.

"Steinmetz resents being called an inventor. He says: 'I am only an engineer. My business is to construct engines that will transport an elemental form of energy into a million factories and homes, dividing this energy up into infinitesimal parts so it can be practically used to run sewing machines, to churn, to wash dishes and to do the dead lift and drudgery that otherwise would have to be done by human hands.'

"So let Steinmetz stand as a type of the modern engineer, who not only is an engineer but is an artist, an economist, a teacher, a humanist."

To have seen Dr. Steinmetz in his office and laboratory in these years of his unceasing usefulness to the world in general would never have given one a picture of spiritless drudgery. Nor would it have presented a spectacle of a man working at feverish tension, nor yet of one who allowed any "unforgiving minute" to slip by without its full "sixty seconds' worth of distance run."

Upon his arrival at the office, he would give his attention at once to whatever was awaiting his advice or assistance. If there was to be a conference, he would go direct to the conference-room, gaging his time so that he would arrive a few minutes or even seconds before the time of the meeting, never allowing himself to be late, unless something unforseen occurred to delay him.

On days when there was no conference, a list of his appointments would be handed him. These he would examine, and then, remarking, "Let me look over the correspondence first," would turn to his mail, answering by making notes in his own short-hand at the headings or in the margins of those letters requiring immediate attention, and handing them to his secretary. The longer papers and more com-plicated matters he took home to study and then to write out the answers, or his comments, to be tran-scribed the following day.

His correspondence disposed of, he would go back to the list of appointments, being prompt and careful in filling all engagements whenever and wherever possible. And he never spared himself if he could be of service to others.

Conferences were usually the occasion for mixing a little pleasure with his work. Often he would throw out a witticism or two, which brought forth hearty laughter. He was fond of fresh air in his office, so that the windows were open to some extent both sum-mer and winter. On one bitterly cold day, when all other offices had their windows tightly closed, the window in Dr. Steinmetz's office was open just far enough to keep the air fresh. A conference was in progress, and the atmosphere eventually became heavy with smoke. One of the men sat directly in line with the window and felt the cold air rather un-comfortably.

"Doctor," he began deferentially, "I think that window——"

"Yes, yes!" exclaimed Dr. Steinmetz, in his quick

way, "I was just thinking that it might be opened a little wider!"

And "a little wider" it was opened forthwith; although when he discovered that his companions were uncomfortably cold, he was prompt to close it down.

If he had no important conference, then, after attending to his correspondence, he would leave the office to follow up tests in the laboratory, saying, "I shall be in the laboratory"; or, if nothing more demanded his attention, he would say: "Well, I guess I'll go home. So-long!"

This was almost always his farewell greeting on every occasion, no matter to whom he was speaking, and no matter whether it was at the office or at his home. Seldom, if ever, was he known to say "Good-by." It was always the more familiar, fraternal "So-long!" spoken with a peculiar inflection of the voice, the first word on a rather high key, and the second lower, with a long intonation, amounting almost to a hum.

No one ever requested an interview or asked him to deliver a lecture and was refused, if it was considered of any importance to the common good. He co-operated with the publicity men who sometimes approached him, whenever it was shown to him that it would materially assist them in their work. And yet Dr. Steinmetz was not in any sense a seeker of publicity. He always had to be convinced that it would really be of constructive value to the electrical cause;

and on many occasions he expressed strong personal reluctance toward engaging in a proposed publicity program, consenting only upon learning that it would advance the work of enlightenment for all who might be benefited by the wonders of electricity.

"Somewhere," one of his associates once remarked, "I have read the words 'everlastingly patient' used to describe the doctor. To my mind no words more fittingly portray him."

Through all the years of his residence in Schenectady, Dr. Steinmetz had gradually made his camp on Viele's Creek his chief delight during the pleasant part of the season. He enlarged it a little at a time, until during the last few years of his life it was a roomy, long, rambling structure consisting of four or five rooms, all joined together, and each giving the appearance of a separate wing.

There he was to be found, with unvarying regularity, from Friday afternoon until Monday morning, throughout the spring, summer, and early autumn, whenever he was in town. As early as the middle of March he would begin his pilgrimages to this enchanted spot, hastening out from his office when the season arrived to make the first trip with as much eagerness as a boy let out for a holiday.

He would find the camp in need of more or less setting in order, cleaning up, and renovating. The canoes would have to be brought out from winter quarters and other duties performed. He would get into his bathing-suit the moment he reached the camp, no matter if it was still the chill of early spring, with huge cakes of ice yet to be seen on the creek and the river from the wide windows of the camp, choking the stream from shore to shore.

Up to the time that the New York State Barge Canal was laid out, making use of the Mohawk River for its route over a considerable distance, the entire Hayden family lived at the camp throughout the summer. It was an ideal spot for the children while they were growing up, with ample room to play on the grassy banks of the creek, plenty of variety in occupation, and not enough water to be dangerous. The men would come out every night from the works and go back in the morning. In those days Dr. Steinmetz made the trip regularly on a bicycle, but for a number of years before his death he had depended largely upon a motor-car that he owned.

After the canal went through, increasing the water in Viele's Creek so that the erstwhile lazy little stream became a wide estuary that compared well in size with the river itself, the camp was considered a bit too dangerous for the Hayden youngsters to stay there continually. Accordingly, they were only taken out now and then on a visit, leaving the doctor and Hayden to enjoy camp life by themselves; and frequently the doctor was there alone. Whatever plans the others made, however, he himself could not be induced to go elsewhere as long as the season lasted.

Steinmetz seated on the famous dam at the camp. With him is Joe Hayden, his oldest adopted grandchild

LIFE IN OFFICE AND CAMP

The water was both broad and deep after the canal project was completed. It came close to the bank on which the camp stood, and completely flooded some of the low land and islands that had formerly been scattered up and down the creek. The site of the famous dam was quite submerged; there was no longer any need of a dam, for the doctor could float off from his little wharf in his one-man canoe and paddle wherever he would without the need of watching for mud-bars.

Those early days were always recalled by Dr. Steinmetz with a vein of whimsical comment. He liked to look back upon the summers when he first knew Hayden, before there was any Wendell Avenue place, and before Hayden had thought of marriage. Dr. Steinmetz alluded to those boyish days, with a jovial smile, by the inclusive expression, "When we were bachelors!"

In the later years of his life, the camp was again the scene of much youthful activity. The two Hayden boys, Joe and Billy, had then grown up to be the boon companions of the doctor. The three of them, an odd, yet picturesque, bathing-suit trio, splashed around on the river and puttered about the camp in a perfect abandonment of content.

Dr. Steinmetz made frequent trips of adventure with Joe Hayden, finding keen delight in exploring the creek and the river, up-stream and down, poking into inlets and recesses for miles in both directions.

The son of his esteemed former employer, Rudolf

Eickemeyer, Jr., was an interested visitor at camp in the summer of 1907. He received a genial letter, dated "Camp Mohawk, August 10, 1907," in Dr. Steinmetz's curiously neat handwriting, in which he was cordially invited to be the doctor's guest. Explicit directions were given him how to reach the camp after arriving in Schenectady, in case Dr. Steinmetz was not able to meet him at the train. Following these instructions, the visitor could not fail to find the camp; "and there," the letter continued, "I hope to keep you until you get tired of camp life. But you must not expect much from a camp. However, I hope to keep you busy with the various camp duties." Then the letter closed with the favorite Steinmetz expression: "Well, so-long; see you next week."

When Mr. Eickemeyer reached the camp, he found it apparently quite deserted. He called out "Steinmetz!" but, receiving no reply, had to repeat the call. Then he heard a faint answering hail, and presently observed Steinmetz approaching from the direction of the river, attired in a bathing-suit, and carefully bending the tree-branches and bushes to one side as he came along the path. He gave his guest an enthusiastic greeting and for several days kept him busy, indeed.

They had some profitable reminiscent talks together, especially in the early evenings, which had begun to be rather chilly. The low temperature caused Mr. Eickemeyer to put on his overcoat and wrap a blanket around him as he lay in a hammock out under the trees. Dr. Steinmetz, however, lounged around in his bathing-suit regardless of the temperature and leaned in characteristic pose on an old drawing-board.

The camp was the scene more than once of an informal conference with the doctor on some pressing matter of interest to the electrical fraternity or to the public at large. Occasionally Dr. Steinmetz was sought after by people who came to see him on business—newspaper men, engineers, fellow-scientists— and it has been the rough, unpainted camp by the river that has witnessed some of these significant meetings.

It is told that a distinguished party of Russian business men and engineers once arrived in Schenectady to consult with Dr. Steinmetz regarding some engineering problems in their own land in the days before the war. To confer with him was one of their principal objects in making the long trip from Russia to America.

It was in the early springtime, just when the lure of outdoor life was beginning to get into the air. The Russian delegation inquired for Dr. Steinmetz and found that he was at his camp, overhauling his boats. So they motored out to the camp, some six miles beyond the city limits of Schenectady.

They were conducted along a riverside path to the door of the low wooden camp building, where they paused, a group of polished gentlemen in tall hats, shiny shoes, and well-tailored suits and coats. In response to a knock, the door was presently opened, and there stood Dr. Steinmetz himself, in an old sweater and a pair of very muddy rubber boots.

He welcomed the delegation with the utmost courtesy and sat down with them at once around one of the camp tables, where they discussed the problem for an hour or more. The contrast in appearance did not ruffle the equanimity of Dr. Steinmetz in the least, nor did it interfere with his cordiality as he finally bade his callers good-by. And whatever the Russian magnates may have thought of the Steinmetz style of camp attire, they seemed thoroughly satisfied with his professional advice, which they had traveled four thousand miles to secure.

Newspaper men occasionally journeyed out to his camp to ask his views on this or that electrical matter, or to secure his comments on some scientific development. Several times they reached the camp only to find it empty at the moment. But after looking up and down the river they would discover the doctor, paddling contentedly about in his small canoe.

"Dr. Steinmetz!" the cry would go forth, across the watery distance. "There's a newspaper man here who would like to talk with you a few minutes if possible."

And the answer would immediately come from the distant canoe, in the familiar thin, high voice of the doctor:

"All right! I'll be there in just a minute."

Then he would proceed to paddle leisurely up to the landing, step out of his dimunitive craft, and accompany his interviewer into the camp, there to answer questions and to talk, perfectly at home, in his well-worn red bathing-suit. His arms, neck, and face were always a deep bronze from the summer sun, and he was the picture of an out-of-doors man in rugged health.

Those who were sufficiently intimate friends to spend a day now and then at the camp with Dr. Steinmetz and the Haydens found it deliciously amusing at first, especially if they were not familiar with the doctor's habits and hobbies. It hardly seemed possible that this easy-living, water-loving, unconventional man, who was to be seen around the camp all day in his bathing-suit, always pleasant, always courteous, was the noted electrical engineer, a world-famous mathematician, and one of the three or four great scientists of his age.

But after a time the visitor invariably realized that Dr. Steinmetz was merely a genuine human man, notwithstanding his genius, and that like any other man he wanted to be comfortable, even if sometimes given to departing from conventions. And so before long the visitor would come to feel perfectly at ease in the presence of a great thinker who was himself at ease, until, when the time arrived to say good-by, and Dr. Steinmetz came forward to shake hands, attired still in the old red bathing-suit, perhaps with his feet perfectly bare, yet always with a bow that was gracious, a manner that was courtly, and a handshake that was sincere, no one thought of his amusing appearance but only of his hearty, gentlemanly warmth of feeling.

"A day in the camp of Charles P. Steinmetz," said a writer who found the sloping, tree-arched, leaf-bedded trail to the plain pine structure one mellow afternoon, "is like a day taken from the life of one of the great philosophers of ancient Greece. He goes out of doors in the summer because he wants to be out of doors. To go out and then put something ·between oneself and the thing one wishes to attain is foolishness, he believes, so he interposes no obstacles between Charles Steinmetz and his friend Dame Nature.

"As a result, there is not a single luxury in the plain camp he built twenty-five years ago—first one room and then another and another as the years went by, until now it is a rambling four-roomed affair.

"There are no carpets on the floors, no pictures on the bare wooden walls. The man whose salary as engineer for the General Electric Company would permit him to have any desire of his heart, gets along with three simple cots, a swing bed, a rough work-table, an alcohol-stove of generous size, some old chairs, a dining-table, and a big cupboard.

"Billy and Joe Hayden, of Schenectady, sons of J. LeRoy Hayden, the adopted son of Steinmetz, live with him at the camp throughout the summer. It is a loosely knit establishment in which every one does something for the common good and no one is fussy.

"Dr. Steinmetz, who retires at nine o'clock every night, is an early riser. He is out and around usually before the boys get up. He cooks breakfast on the alcohol-stove, just as he does all the rest of the meals. He is a champion pancake maker. Charles P. Steinmetz gets as much fun out of flipping a flapjack as he does in turning an algebraic formula into something that means electrically less steps for millions of persons. He bakes them big and round, one at a time, and enjoys eating them as much as he does making them.

"After breakfast he goes to his work. If the day is rainy, he works inside at the rough table he made for the purpose long ago; but at all other times he goes to his office on the Mohawk.

"Probably there is no other office like it in all the world—a battered twelve-foot tippy canoe with a cushion in the bottom and four boards laid together from gunwale to gunwale, thwartwise to serve as a desk.

"When he goes down to the river to work he carries his papers under his arm, with them Hutchinson's volume of four-place tables, and a little Nabisco box wherein he keeps his pencils.

"Depositing these in his canoe, he pushes off from shore, gives a few vigorous strokes with his copperbound double paddle, and prepares for the morning's work.

"He lays the four boards together carefully, and with precision places his papers on them, each pile weighted down by one of an assortment of pebbles he carries in the bottom of the canoe. If the day is too warm, he slips off his shirt and, putting it in the bottom of the canoe, works in his undershirt.

"Through the long summer morning, the light winds of the Mohawk push his craft up and down the river, and for hours at a time he bends over his improvised desk, all unheeding."

It was sometimes said that Charles P. Steinmetz could read anything that interested him. He was found one day out in his canoe absorbed in a book that had no cover. One glance was sufficient to show that it was not a technical book; he was not at work this time but reveling for the moment in complete relaxation.

His visitor ventured to inquire the title of this work which had cast such a pleasant spell over the master of electricity. Dr. Steinmetz smiled broadly but shook his head.

"It's a book called 'The Lunatic at Large,'" he answered. "I don't remember the author's name. The cover got torn off somehow. But it is interesting."

The reply was typical of the man. He found recreation very often in light fiction, sometimes in exciting novels, or detective stories. Yet his serious tastes in reading were broad and sound. He was once asked to name a list of ten books which he would prefer to have beside him, of all other books, should he be compelled to pass the rest of his life on a desert island. This is the list he drew up: Homer's Odyssey, the Odes of Horace, Goethe's "Faust," Mommsen's "History of Rome," Kipling's "Jungle Book," Mark Twain's "Tom Sawyer" and "Huckleberry Finn," Stevenson's "Treasure Island," Marryat's "Midshipman Easy," Stanley's "In Darkest Africa," and Sienkiewicz's "The Deluge."

This list was published as Dr. Steinmetz's group of "books I have enjoyed most"; and shortly afterward a writer in the "Springfield Republican" commented on his selection in these words:

"From the classical languages come the greatest of tales of adventure, the Odyssey of Homer, and the most polished of vers de société, the Odes of Horace. German literature contributes its one indisputable masterpiece, Goethe's Faust, and the Roman History of Germany's greatest historian, Mommsen. For historical fiction he turns not to Sir Walter Scott but to the Polish novelist, Sienkiewicz, whose romance, The Deluge, is one of the most instructive books about Europe that a modern reader can find; it shows vividly the struggle with the Tartar hordes on the eastern marshes.

"English literature is represented on the list by Darkest Africa, a record of romantic exploration which is still fascinating in spite of the changes since Stanley's day, and by four works of fiction, all of

which would be called light—Marryat's Midshipman
Easy, a classic picture of the old-time English navy;
Mark Twain's companion books, Tom Sawyer and
Huckleberry Finn, the only American literature included; R. L. Stevenson's Treasure Island, the
polished jewel of literature for boys, and Kipling's
Jungle Books, a work on which their author's reputation could well rest. Anybody wrecked on a desert
island with the ten favorite books of Mr. Steinmetz
would be well provided for."

This was the period of Steinmetz's career when he
was at his zenith. Mature in a peculiarly vivid, profound sense, recognized the world over for his astute
judgment and his intensive knowledge in several
fields of human interest, listened to with respect when
he spoke and consulted by high and low in his own
profession, he placidly moved through the halcyon
days of his life without the slightest evidence of conscious superiority or deliberate ostentation.

It' was in this period that he computed the probable value of a lightning-flash; discussed the possibilities of radio, especially of radio transmission of
power; indicated his increasing interest in the entrance of electricity into the home as a tireless assistant to the modern housewife; and looked back with
fond retrospect upon the start in life which he
credited America with having given him.

"I love America," he said. "It took me, a crippled
boy, and gave me a chance. I have faith in it, and,
through my electrical work, I want to help the

America of the future. We will succeed the soonest by giving every one a chance. But we must do things in the most economical way."

Several of his most interesting encounters with other scientific notables occurred about this time. In 1921 he was one of those who welcomed Professor Einstein to America, and went with him to inspect several radio stations in eastern states. Dr. Steinmetz was generally supposed to be one of the dozen especially favored individuals who were able to comprehend the famous Einstein theory of relativity; indeed, he published one or two articles about this time in which he discussed the Einstein theory in somewhat popular style.

In 1922, Signor Gugliemo Marconi, pioneer of radio development, met Steinmetz at Schenectady after an interval of about fifteen years since he had last seen the electrical master. They were not together long, but long enough for Signor Marconi, who was well acquainted with some of the Steinmetz hobbies, to inquire solicitously after the doctor's pet alligator and Gila monster, neither of which, he was gravely informed, was any longer a protégé of Dr. Steinmetz.

It was in this same year that Dr. Steinmetz and Thomas A. Edison met for the last time. The noted inventor of East Orange had come to schenectady to see what was going on at the General Electric works. Part of his interesting day's round of inspections was devoted to the general engineering laboratory, where

CHARLES PROTEUS STEINMETZ

Dr. Steinmetz conducted his experiments with artificial lightning.

Dr. Steinmetz carried out a demonstration of his lightning machine for Edison's benefit, causing the lightning to play pranks with porcelain insulators and with the limbs of small trees. Then, while cameras clicked busily, he and Edison examined the shattered fragments of these objects, while Steinmetz explained what had happened and its significance.

It was toward the close of this part of his life, too, that he gave several popular addresses by radio, speaking from the studio of WGY, the broadcasting station of the General Electric Company at Schenectady. The first of these was a talk on "Lightning," in which he explained the source of lightning and estimated the probable voltage of a lightning-flash.

In another radio address he rather astounded many of his hearers by declaring that he doubted the generally accepted theory of the existence of the ether. He frankly did more than doubt; he frankly asserted his opinion that there is no ether and that space is—just space. This view of his was subjected to some criticism, especially as he did not undertake to state the whys and wherefores except in somewhat incidental fashion. But he did not urge the theory. Never, so far as is known, spoke of it in public again, nor pressed it with any emphasis in private conversation.

The approach to Camp Mohawk in 1913. The camp had by this time been enlarged until it consisted of several rooms, joined together

The camp in later years. This shows how the structure looked after all the additions had been made. Dr. Steinmetz is standing with the two Hayden boys in the original portion

ANDREW BALET/2006

CAMP MOHAWK – 2006

LIFE IN OFFICE AND CAMP

Throughout these years when he was known and looked up to by numerous men in his own profession, and regarded with admiring awe by the public at large, he was called upon at intervals for after-dinner speeches. He gave a few of these, in some of which he enunciated some of his best-remembered predictions in economics and physics. He had a broad strain of humor and could usually make his audience laugh with regularity. This ability, however, was somewhat restricted by his peculiar thin, high voice, with just a trace of accent; although throughout the last thirty years of his life he displayed a perfect mastery of English.

In the midst of these busy, useful days, he was called upon to serve at the head of two national bodies which honored him for his service in two distinct departments of human endeavor. The first of these elevations was his election in 1915 as president of the National Association of Corporation Schools; the second was his designation for the same position in the Illuminating Engineering Society for the year 1915–16.

CHAPTER XXIII

PERSONAL WHIMS AND TRAITS

LIKE Eugene Field, who continually amazed
and amused his friends by his eccentricities
of daily life, Dr. Steinmetz was concerned
more with "affairs of the mind and the heart," than
with conventionalities of society and dress. His was
a simple, democratic disposition, asking little of
others in the way of formalities and rather taking
it for granted that others would ask little of
him.

There was not the slightest suggestion of discourtesy in this characteristic attitude. Few men
possessed more natural courtesy or consideration for
others than did Steinmetz. But the niceties of eti-
quette and changing styles in clothes simply did not
appear to him to be of half so much importance as
the solution of a complicated problem in higher
mathematics, or a computation with logarithms by
means of which the usefulness of electricity to all
mankind might be increased.

Hence, he devoted most of his waking hours to
these intensive efforts and was unwilling to take the
time to find out about matters that impressed him
as unimportant.

PERSONAL WHIMS AND TRAITS

No one who ever became a close friend of Steinmetz's thought any the less of him because of this disposition. The personality of this wholly human scientist was extremely charming. He radiated an atmosphere of kindly companionship, coupled with extraordinary keenness of mind and quickness of perception; and it is difficult to say which of these elements was wont to command the greater admiration among his friends.

He was undeniably a man of hobbies. His fondness for plants of the desert was unique. He delighted in flowers, and his grounds on Wendell Avenue included a luxuriant flower garden. He loved Nature in all her moods, and loved all of Nature's works.

There is a whimsical story of how his delight in woods and plants and growing things reacted to his pecuniary advantage when he purchased his property on Wendell Avenue. The land included a clump of woods adjoining the little dell, through which runs a small stream. Supposing that the average land-owner would buy simply to build and would be apt to regard this grove as a liability on the property, the agents for the land allowed a deduction in the price on account of the wooded stretch. Dr. Steinmetz said nothing at the time, but after the transaction was closed he confided to Hayden that he considered the little grove on the property one of the best things about it; he was delighted to have it there, feeling fearful only that the agents would discover how much he would enjoy it—and consequently increase the price instead of reducing it!

His hobby for pets has already been referred to; likewise his fondness for collecting. His camp and the opportunity it afforded for paddling around the river in a little canoe, only large enough for himself, was unquestionably a hobby. And he was an enthusiastic amateur photographer. He took photographs more or less regularly for twenty years or more, and had hundreds of negatives, all carefully numbered, titled, and signed with his initials. Most of the pictures are scenes at camp or groups of the Hayden family, particularly the two elder children, Joe and Marjorie, with some later views of Billy Hayden, the youngest of the three.

Of all the traits in the character of Dr. Steinmetz —a character of many elements, certainly suggesting his student designation of Proteus to those who saw him in all his moods—none is so winsome as that which reveals him as the comrade of children. Almost all the small boys and girls of the neighborhood knew him personally, and they were sure to receive a quick, cheery greeting from him if they met on the street.

The relations that existed between Dr. Steinmetz and his adopted family were a perfect example of simple, sincere domestic affection. The Hayden children never thought of him as an adopted kinsman; it was always just plain "grand-daddy" with them.

And he was a grand-daddy who was never too busy to listen to their childish stories nor to explain things they wanted to know about. He always found time to help them with a school problem, showing them, with gentle, kindly solicitude, the principles of arithmetic, and exhibiting the while the utmost patience. He observed all their ways and manners, watched how they grasped ideas, and listened, with his quiet smile, to their lively, voluble comments.

Charles W. Wood, who saw much of Steinmetz at one time, both at the camp and at home, records the pleasure Steinmetz found in the study of children.

"It happens," writes Mr. Wood, in the "National Pictorial Weekly," "that this acknowledged head of his profession in America was one of the most human of men. His fondness for children was pronounced. He was as enthusiastic in helping them understand how two and two must be four as he was in demonstrating his latest electrical calculations to a convention of grey-bearded professors. Steinmetz in his laboratory, Steinmetz working out a problem upon which our industrial civilization waits, was no more absorbed in study than was this same Steinmetz, with a handful of fortunate children around him, as he demonstrated the miracle of long division. Looking at him upon such an occasion, it was impossible to tell who was the pupil and who the teacher. For the wizard was learning. He was asking questions. He was finding out how the marvelous human forces operate."

Thus it can be seen that Dr. Steinmetz was a rare grand-daddy, who, ever since the children could remember, had always been there with his understanding sympathy, his quiet, whimsical methods of amusing them.

Dr. Steinmetz at home, when thoroughly at leisure, was almost sure to be found in the company of Mr. Hayden's children while the latter were growing up. He was almost shy toward them when they were very little, but as they reached adolescence his interest in them knew no bounds; and all three of them, Joe, Marjorie, and Billy Hayden, remember many a jovial prank they had together.

Often, on a quiet summer evening, Dr. Steinmetz was part of the family audience at high doings out in the big, picturesque garden, half cultivated, half natural. He would find a favorite spot on one of the little benches beneath the tall pines, and there would rest amid the coolness of the woods—those same woods which were supposed to be an undesirable addition to the property—while he watched whatever was going forward on the green lawn that stretches between the lilac-trees and the borders of the Canterbury bells.

As likely as not it might be an exciting amateur circus which was in progress on that lawn, with bareback riding, clowns, trained animals, and acrobats. Or a Wild West round-up. Or a party of small girls of the neighborhood, whom Marjorie was entertaining. Whatever it was, the doctor was a smiling onlooker, and some times an enthusiastic promoter or an arch coöperator in some gigantic adventure which the children were undertaking.

Throughout their school-days, the Hayden children were absorbing a small treasure-house of information from Dr. Steinmetz. He could always hold their close attention by telling them stories from history, both European and American, ancient and modern; or by drawing upon his inexhaustible storehouse of knowledge to explain to them some of the wonders of the physical world.

Sometimes when Mr. and Mrs. Hayden returned from an evening down town they would notice a light still burning in the children's room—long after hours! Upon quietly going up to investigate, they would come upon an odd scene. Dr. Steinmetz would be there, leaning in characteristic attitude on the corner of the bureau, relating some absorbing narrative to the children, who would be curled up on the bed before him.

And then, some days later, one of the children would bring out a surprising bit of knowledge that would lead his father to exclaim, "Billy, where did you learn that?"

To which Billy would reply, "Oh, grand-daddy told me."

When the children began to advance in their school-work, the doctor followed their progress with unflagging interest. When Joe Hayden went away to preparatory school, the doctor kept up a steady correspondence with him, much of which was written, just for the fun of it, in code, by using the Morse system of dots and dashes. Joe originally suggested the code method, and it appealed to Dr. Steinmetz at the moment, although it must be remarked that the doctor was never a voluminous letter-writer and left but a meager personal correspondence.

The Haydens were a family whose children had the exclusive services of a skilled chemist, a noted scientist, in preparing their Fourth of July fireworks! Did ever boys and girls find so wholehearted a playmate in a man whose fame stretched from one coast to the other, and across the oceans to other lands?

For picture to yourself the interior of a large laboratory room, a room in Dr. Steinmetz's laboratory, adjoining his home on Wendell Avenue, in Schenectady. It is the chemical room of the laboratory, which was added some years after the original building was put up. There is a fascinating confusion of chemicals and chemists' paraphernalia on every hand —tubes and bottles, glasses and burners, and countless rows of glass jars in glass-incased shelves around the sides of the room.

In the midst of it you find Dr. Steinmetz and small Billy Hayden, their heads just as close together as they can get. The two are so absorbed in watching a vial cooking over a Bunsen burner that they do not hear your step at the door. Something is going forward that is weird, romantic, altogether exciting! You sense the tenseness in the air.

You make a sound, and they both look up and smile. Billy, revealing that marvelous sight, a perfectly happy boy, gives you one glance and then goes back instantly to watching the mysterious something that is a-brewing. The doctor, hardly less delighted, pauses long enough to remark, with twinkly eyes:

"We are making fireworks for the Fourth of July!"

Fireworks for the Fourth of July! A whimsical, fantastic occupation, surely! An odd type of relaxation for a man of vast scientific attainments. But, you think at once, it is not so important that the doctor cannot interrupt it to talk to callers. Of course he can postpone this curious laboratory enterprise, whose motive is simply the amusement of children.

Interrupt? Postpone? The mere suggestion of such a proceeding brings a fresh revelation of the whimsicality of the man. Nothing whatever, you discover, can possibly be allowed to interrupt that laboratory business that evening. Callers must wait—or come some other time! Billy and the fireworks are the supreme interest of the moment. Rather than disappoint the children, Dr. Steinmetz, you discern, would ask the most eminent caller to wait.

So the making of the fireworks continues. It continues all that evening. It continues past Billy's bedtime. It is carried on by Dr. Steinmetz alone off there in the laboratory after Billy has gone into the house. It is still a-brewing when the night is far advanced. The doctor stays there by the bubbling retort until the preparation is completed and the novel Fourth of July celebration is assured.

The fireworks—"harmless fireworks" they were called, and harmless they proved—were displayed with great jubilation on the night of the Fourth, at the camp on the creek. They were enjoyed not only by Billy and Joe and Marjorie but also by many campers up and down the river.

And if there was a party for the children at any time, Dr. Steinmetz was likely to find himself one of the invited guests. He would make it a point to be on hand, too; he would no more disappoint the children, when he knew they expected him, than he would the president of the General Electric Company.

On one of these happy occasions, Marjorie Hayden was host to a group of little girls at a table all dainty with pink decorations. In the midst of the merriment, Dr. Steinmetz came quietly into the room from his office in the museum and stood near the door a moment, looking at the bright scene before him.

Presently one of the little girls, a frequent visitor at the house and therefore familiar to the doctor, discovered him and called out a happy "Hello!" And then they all discovered him, and Marjorie had to introduce him to all who did not know him.

She escorted him around the table until he had stopped at every place and spoken to every little girl. He was told each one's name and was greeted with delighted laughter by those who were already acquainted with him.

Dr. Steinmetz's birthday was never allowed to pass unobserved. Billy and Marjorie would not hear of it. They always had their own birthday parties, and they were quite insistent, whenever April 9 came around, that grand-daddy should have one, too. And so, for many years, it was the custom.

These gracefully pleasant occasions can best be comprehended by a description of one of them. It came on a Sunday, so that there was ample time for all the preparations.

Of course the birthday supper was to be a surprise. At least, it was called a surprise, although the doctor knew perfectly well what to expect. The children were fairly a-quiver with excitement, none the less, and upon the arrival of a friend of the family to offer birthday greetings to the doctor, they insisted on helping him exhibit his gifts—especially the things they had presented to him.

Dr. Steinmetz was found in his museum office, examining a handsome edition of a rare old book. Billy and Marjorie and Joseph brought the caller in, and then gathered every one around to see their gift, a brand-new pencil-sharpener. Marjorie and Billy climbed up on top of the desk to see all that happened, while Joseph operated the sharpener. Dr. Steinmetz looked on with twinkling, fun-loving eyes, although he was clearly interested more in the children than in their gift.

The excitement increased as supper-time drew near. The whole family joined in preparing the surprise. What happened when, at length, everything was ready is best told in the words of the guest:

"Mrs. Hayden shooed us all out into the hall. The doors which led to the dining-room were shut, and Marjorie and Billy brought grand-daddy right up just as close as they could get, all ready to rush in when the signal was given. They clung to his arms and danced up and down in their delight, and at a word from their mother sang the kindergarten birthday song, 'Happy birthday to you.'

"Then the doors were opened, and we all exclaimed with pleasure at the pretty sight. The room was lighted with only the birthday candles, which surrounded the cake in the center of the table. The soft rays shone on the charmingly appointed table and more dimly lighted the rest of the room, which seemed half filled with flowering plants. One of the windows looking into the conservatory was open, and the new orchid which the family had just given Dr. Steinmetz hung there, making the air of the dining-room sweet with its wonderful perfume.

"It was a pleasant supper-party, and when we

were ready for dessert, Marjorie said, looking affectionately at Dr. Steinmetz:

" 'The Birthday Child must cut the cake.'

"And so Dr. Steinmetz cut the cake, with the pretty icing on it, which read 'Happy Birthday,' and showed the years of his age underneath."

And his garden—and his pets! Dr. Steinmetz enjoyed nothing so much, in the spring and summer, as an opportunity to show a visitor his garden. It was realy a landscape-garden, to which the wooded patch in the rear of the grounds, thickly grown with pine-trees and bushy undergrowth, gave a picturesque setting.

His garden, of all gardens, was not a mere formal estate for stylish exhibition. It was a place of deep enjoyment for its owner from earliest spring until latest autumn. He liked nothing better than to join the family conference that always occurred when the seed catalogues arrived. He took a whole-hearted relish in the planting and cultivating, supervising with zest, and often putting seeds or plants into the ground with his own hands.

Dr. Steinmetz never failed to get out in the garden as soon as the very earliest blooms began to come out. That would be often before the snow was off the ground, when the crocuses were just appearing around the greenhouse walks and later the snowdrops, wood-lilies, and daffodils blossomed beneath the hawthorn hedges. Later still the tulips made a gorgeous display in the big bed in the front garden. As each new arrival appeared in its proper season, Dr. Steinmetz welcomed it in his own characteristic manner of pleased, absorbed enthusiasm.

His pride in the garden came quaintly to the front whenever there was an opportunity to take a friend around through the pleasant walks. Mrs. Hayden suggested one early spring evening that a visitor who had just arrived would perhaps like to see the lady-slippers. The family quietly got into their wraps and were soon ready to start. At that moment Dr. Steinmetz appeared, smilingly eager to show off his garden, and he headed the procession that passed down by the laboratory and out to the edge of the pine woods.

Through long, winding paths the party proceeded until in a green opening under the trees he pointed out the daffodils, just catching the light of the low sun, and the narcissuses, in pure white clusters. He showed them the strawberry blossoms, the wild lilies-of-the-valley, and the great white trillium with its wondrously sweet scent. Painted trilliums were there, too, almost invisible in the dusk, because of their dark blossoms.

It was a delightful exhibit, delightfully exhibited. And the most refreshing human touch was contributed by small Billy Hayden, who found it tremendously difficult to refrain from picking all the flowers he saw and presenting them to the visitor. Billy and Dr. Steinmetz disagreed about picking flowers. Dr. Steinmetz declared that blossoms should not be picked. He would much rather take his friends into the garden to see the beauties of the flowers. He averred that this allowed them to be observed in their proper surroundings.

But there was a deeper reason, too, and Billy, it turned out, knew what that other reason was.

"Mother and grand-daddy don't want me to pick the flowers," he announced, "so that the blossoms will turn into seeds and fall on the ground and more flowers will come up next year. That 's what grand-daddy says."

The promenade continued, down toward a little stream that runs through the end of the garden woods only in the springtime. In its damp vicinity luxuriant ferns and wild orchids were growing, as well as groups of pitcher-plants, which were sent to Dr. Steinmetz from the Adirondacks. And there the lady-slippers were discovered, blossoming for the first time.

"They looked," wrote the visitor, describing the scene, "as if Nature herself, and no mere gardener, had planted them, quaint little orchid things with their long twisted petals, like the golden curls of a fairy.

"Then down by the fence we went, to see how the cowslips were growing, and up along the farther path to where a new wild azalea bush, dug up in the woods the Sunday afternoon before, was looking at home in its new place. The laurel, brought from the Berkshires, the rhododendrons, from the prosaic florist, seen here and there under the pine-trees, were all getting ready to blossom a little later, to make the woods lovely with their beauty after the earlier spring flowers had gone.

"We admired Billy's and Marjorie's flower-beds, which had just been planted, and the tiny pond which the stream from the conservatory made, surrounded by a mass of blue and gold iris and large, sky-blue forget-me-nots."

That is a glimpse of Dr. Steinmetz among his flowers and in his garden. Scenes equally captivating, if somewhat more grotesque, often were to be witnessed when he was among his pets.

A good many of the pets that Dr. Steinmetz owned in later years had been presented to him. These included some of the queerest he ever had. Among them was his famous Gila monster, which was kept in a roomy cage in his laboratory. When he first owned it he allowed it to wander at will in the conservatory, but persons who encountered it unexpectedly were so badly frightened that he finally had the cage built. It was a sluggish thing, sleeping most of the time, and finally dying from decreasing interest in life. Dr. Steinmetz said it was a case of lack of ambition and loss of appetite.

Gila monsters, Dr. Steinmetz had heard, can inflict deadly poison by their bite. He endeavored to verify this by making inquiries among zoölogists, but the opinions that were advanced differed so much that he decided to see if he could find out for himself.

Accordingly he bought some rabbits with which to make the experiment. But when he got the rabbits home, they were such friendly, interesting little chaps that the doctor could not bring himself to sacrifice them, not even for the sake of the desired knowledge. And so they were added to the Steinmetz family of pets, where they were apparently perfectly happy, although eventually, in a passing whim of ingratitude, they ran away.

Dr. Steinmetz brought three parrots with him when he moved into his Wendell Avenue home. One was green, the second was white, and the third was gray. They were all handsome and unusually tame, at least with Dr. Steinmetz. But they were great chatterers, and when they were not talking they were screaming. He finally had to separate them from one another on account of the noise. He did this by putting one on each floor of the house. Each was at liberty to roam about on its own floor but was not allowed to invade the other floors. In that way they were kept more quiet, and it was possible to live in the house without being irritated by a continual hubbub.

Baby alligators were more than once serene inhabitants of the pool in the orchid house; tropical bugs and a good-sized school of goldfish were also to be seen there. The doctor once had a couple of raccoons, which he kept in a cage over the brook until they broke loose and scampered off. Several odd varieties of owls had also been presented to him at different times.

As for kittens, Dr. Steinmetz supplied the family with one or two delightfully amusing incidents, resulting from his dealings with that species of the dumb creation. A group of men were passing through the home laboratory one night when they caught sight of a wooden cage with straw in the bottom. It contained a small animal of some sort. Recalling the many stories that were told of strange pets owned by Dr. Steinmetz, they inquired, when they had opportunity, "What have you got there in the cage, doctor?"

"Oh," answered Dr. Steinmetz, with a look of mock seriousness, "that's Marjorie's kitten."

And so it was. The cage, it seemed, had once been the "cradle" of a puppy, which had been graduated from its training quarters, and the children had decided it would be just the place for the kitten to sleep.

Not long afterward one of the children brought home another kitten, a homeless waif. The kitten that enjoyed the luxury of the cage resented the presence of the new-comer. Soon the two were sworn enemies, spending most of their time spitting at each other through the bars of the cage.

Dr. Steinmetz observed this situation with grave concern. At length he assumed the rôle of disciplinarian and undertook to teach the kittens a code of ethics. In the exercise of this commendable crusade for feline uplift, he was one day discovered down on his knees by the cage with a long willow branch in his hand, with which he was gently whipping the more turbulent of the two kittens and saying in a coaxing tone:

"No, no, catty; no, no!"

But the kittens continued as refractory as ever, and the difficulty had to be finally solved by giving one of the little "catties" away.

The four-footed, feathered, and crawling friends of Dr. Steinmetz were somewhat depleted in latter days. Through one circumstance or another, their number decreased. Dr. Steinmetz summed up the vicissitudes which have more or less surrounded them when he told Marconi, in reply to a question which the latter asked when the two famous scientists met in the summer of 1922 at the General Electric plant:

"Oh, I have n't many pets now. I have had to give some of my pet alligators away. And the Gila monster died. He was too lazy to eat."

The people of Schenectady came to be accustomed to seeing Dr. Steinmetz going about without a hat during all except the three or four coldest months in the year. And during these months he would appear with a coon-skin cap, having large ear-flaps. No one could ever recall having seen him in an overcoat, regardless of the weather. Indeed, his disregard for winter wraps caused anxiety among the Haydens, who endeavored to show him that his health was at stake. But the best they could do was to effect a compromise by inducing him to put on extra heavy underwear.

He was far too democratic to ride in his own automobile, although he had a motor-car which he used in later years to get out to his camp; in his earlier life he made the journey to and from camp on a bicycle. In going to and from his office at the General Electric Company, however, he commonly rode in the street-cars, invariably going to the very front of the car, bidding a cheery good morning to the motorman, and then intently watching the rails as the car sped along. Even if he were talking to some one who accompanied him, he would keep looking ahead down the track, as though fascinated by the way the car moved over the ground—a contrast, indeed, to the discouraging efforts to run electric cars in Brooklyn, of which he was a spectator many years before, in the time of Eickemeyer and Field.

So irksome did he find it to give attention to clothes that many amusing stories got into circulation concerning his sartorial habits. It is quite true that he hated to buy clothes—Hayden usually did all the choosing of materials for him—and that only exceptional circumstances could induce him to appear in conventional dress. He usually wore a neat business suit to the office, but was as likely as not to have on a sweater under his coat instead of a shirt, or to

Steinmetz seated on the famous dam on Viele's Creek, toward which all week-end guests at the camp were expected to contribute their labor

forego the small matter of a tie. And no one ever remembered having seen him in a dress-suit.

Once in a while, on rare occasions, he was induced to wear a somewhat stylish suit of clothes, with high collar and tie, for the purpose of being photographed. This was very seldom, however; and his well-known indifference to clothes always caused recurring speculation by his friends as to just what might be the extent of his wardrobe.

Once or twice he quite forgot appearances during the absorbing interest of an address that he chanced to be delivering. Mr. Hayden recalls one such occasion, when they were in Chicago and the doctor was to speak before the American Institute of Electrical Engineers. The meeting was held in a room that had been used as a lodge-room, and on the platform were several big chairs, upholstered in red velvet. The platform was high and the hall large and well filled. Dr. Steinmetz was standing near one of the chairs and thoughtlessly put one foot upon it. It occurred to him that he could see everybody and make them hear better if he were a bit higher up, and so he stepped upon the chair. There he stood for the rest of the speech, and Mr. Hayden, down in the audience, saw with horror that he had forgotten to take off his very muddy overshoes.

"So I usually traveled with him after that when he went to make a speech, to remind him to take off his overshoes. Of course, nothing could keep him from climbing on the chairs if he wanted to," said Mr. Hayden.

But he was not quite as indifferent to his personal appearance as some of the magazine writers have tried to make him appear. There was a story in a magazine once which said that Dr. Steinmetz never bought a new suit until the one he was wearing was fairly in rags. Just as the family were teasing him about it, although they were a bit provoked at the time, the postman came, bringing, among other things, a bill for one hundred dollars from Dr. Steinmetz's tailor for his winter suits, which had recently been sent home. But Mr. Hayden always chose the materials, and Dr. Steinmetz refused to give up any time to visit the tailor for fittings. The tailor always came to the house.

With all these idiosyncrasies, there was never any question as to the simple sincerity of Dr. Steinmetz in all his views, nor of the bigness of his heart. He had a childlike faith in mankind and in the goodness of human beings. During the European War his distress over the unloosing of human passion and wantonness was far deeper than was his concern over the political issues.

The war saddened him, without regard to the side that seemed to have the advantage at the moment. It saddened him still more when the United States joined in the conflict, for he then found that, notwithstanding his twenty-three years of American citizenship and his unswerving allegiance to American ideals and institutions—his complete harmony, in fact, with the spirit of America—his position was a difficult one.

But his conviction that, even in war-time, men's motives were not wholly diabolical remained unshaken; so much so that he was found one evening displaying some dumdum bullets that some one had sent him, and affirming with the utmost earnestness that surely neither side meant actually to use bullets of that sort in battle. They had simply been hastily gathered up without any definite purpose; but, assuredly, they were never intended for the firing-iine!

Such was the gentle trust of this most human of scientists that his warm heart was incapable of believing there could be such a thing as twentieth-century barbarism in his fellow-men.

CHAPTER XXIV

WHAT STEINMETZ THOUGHT OF RELIGION

"THOU servest," might well have been inscribed in imperishable letters after the name of Charles P. Steinmetz from the moment of his maturity. For his was a life that served his fellowmen, by achievement and direct intent; and not in appearance only, but in golden reality. His was an altruism perfectly unostentatious. There was no trumpeting abroad, no words without deeds—but useful deeds and few words.

In the self-forgetful realm of practical service to the world at large and everybody in it, Steinmetz assuredly stood close beside Leigh Hunt's embodiment of the unselfish character, his immortal *Abou ben Adhem*. Like *Abou,* Steinmetz loved his fellowmen; and he did much for them.

Some would put this down as the real religion of Steinmetz, the religion of service, which, also, was taught by the man of Nazareth. Needless to say, it was a far finer thing, this ideal that governed the life of such a scientist, than the dogma-bound, creed-canting attitude that is lacking in real works. Steinmetz would have none of creeds, and no man was less dogmatic, even in the realm in which he was a master.

444

STEINMETZ AND RELIGION

He had made it evident, at one time or another, that he could not be regarded as other than an unbeliever, in respect to the Christian faith, until to those who knew him and had long had the opportunity to grow familiar with his views he had come to be definitely looked upon as an agnostic. Some might go further than this and describe him even as an atheist, but it is by no means certain that he did not hold a belief in God. He had given indications of believing in a Supreme Being.

From his public utterances and writings, however, the impression is clearly derived that Dr. Steinmetz's mind had become so steeped in the methods and point of view of science that he could not approach the big question of the existence of a spiritual side in man except by the methods that science usually pursues; and that meant, for him, the impossibility of believing what cannot be logically proved.

However much Dr. Steinmetz practised altruistic service and lived under the dictates of human sympathy, however big-hearted and sociable he was, however winsome and gentle in character as his career developed, he was unable to cultivate in the higher sense the one essential to the religious life, faith.

He himself recognized and designated faith as the basic requirement in religion. And he failed to find that faith in his own heart.

Dr. Steinmetz, for many years during his later life, maintained the friendliest relations with a society of the Unitarian Church in Schenectady. It was All Souls' Church, the church in which Mr. and Mrs. Hayden and their children were active. That circumstance accounted for his own interest in the organization, and at the same time makes clear the extent to which he entered into the pursuits of the Hayden family. What they were interested in, he was interested in, and where they went, he went. True, he did not make a practice of attending service at the Unitarian Church. But he was present on a number of occasions when there were events participated in by the Hayden children. Comrade of the children as he always was, Dr. Steinmetz would sacrifice time and convenience to attend their various merry gatherings. And when a Sunday-school event occurred at the church, he was frequently to be found in the audience.

He became a friend of the Unitarian minister at All Souls', the Rev. Ernest Caldecott, with whom, on several occasions, he discussed the fundamentals of religion. Mr. Caldecott found him anything but antagonistic toward the church, even though he did not believe in the faith which the church represents. But these conversations between the two also uncovered the final attitude of Dr. Steinmetz upon religious faith, his lack of belief, his agnosticism.

To Mr. Caldecott, more than to any other one person, he imparted his inner convictions upon this subject. Mr. Caldecott has set them down in what may be called as close an approach to an analysis of Dr.

STEINMETZ AND RELIGION

Steinmetz's personal attitude on religion as can be obtained. This is what Mr. Caldecott has to say about Steinmetz and religion:

"If religion consists in setting a high value upon life and in being loyal to that value, then Dr. Steinmetz was an intensely religious man. If it consists in holding certain theological views, then he was not religious. Everything depends upon one's definition.

"Steinmetz accepted the general understanding of the term, for he had stated that 'religion deals with the relations of man with the supernatural, with God and immortality, with the soul, our personality and the ego, and its existence or non-existence after death.' This he called 'the greatest and deepest problem which ever confronted mankind.'

"His views were practically negative on the entire problem; he was simply an unbeliever. Having given serious consideration to the matter he had decided that there was nothing for him to believe. This has placed him before the public as an atheist. The title he did not deny. The writer, however, would put him down as a confirmed agnostic, for an atheist is a person who knows there is no God, and Steinmetz was not of that temperament. Be that as it may, the fact remains that he was a consistent unbeliever.

"How much of his thinking upon the question was influenced by his own experience of life will probably never be known. But it seems likely that his inability to do as physically normal men do had an effect upon his religious ideas. With that penetrating mind for which he was famous, he saw reason to believe that no almighty goodness ruled this universe. A malicious great being was unthinkable. All was according to cause and effect, a process which had wonderfully evolved, without personal direction or plan, an amazing series of adaptations, with man but an incident in the whole.

"His analytical mind was kept balanced by genuine sympathy, so that no bitterness was present in his unbelief. His only complaint against religion or the church, he once told the writer, was his strong suspicion that the clergy were afraid to tell the truth as they saw it.

"This background is necessary in order that one may appreciate the views on religion held by Dr. Steinmetz. On the side of theory he asserted that religion dealt with infinity and science with finite matters. Obviously, they were two separate and distinct fields. In the nature of the case, there could be no conflict between the two, unless one or the other crossed over out of its own territory.

"Religion, he declared, is based on an assumption which cannot be proved. Time and again religious exponents have been compelled to shift their ground because science has entered and given explanations of phenomena which had previously been interpreted in terms of supernatural intervention. Room seems not to be even left for the existence of a 'vital force,' for one after another of the chemical products of living things have been synthetically produced by chemists, until it seems but a matter of time before the very heart of the universe will be reached.

"All this, said Steinmetz, would be undisturbing to man but for the conceit of his own ego in the idea that he himself is immortal and needs God in token of that immortality. Yet in the mind of the scientist the same principle obtained for man's mind as for supernatural forces, viz., new discoveries tend ever more strongly to show that life is a physicochemical process and that when this process ceases mind ceases. Thus man believes in his own continuance after death because he wants to, and not because there is either reason or evidence that continuity is a reality.

"It would seem that Dr. Steinmetz regarded the non-existence of God as a supernatural being as automatically excluding immortality. The writer had arranged to discuss this matter with Dr. Steinmetz and was only prevented from doing so by the latter's sudden illness and death. But, without making any plea for continuity of existence, it would surely appear reasonable that the combination of forces through which all that is has been brought about, with man as its apex, might, conceivably, be sufficient to make possible a continuance of consciousness after mortality. At any rate the non-existence of God does not exclude the possibility of immortality (to use the old word, which is an unfortunate one). If man is not immortal it is on grounds other than that

God does not exist except in men's minds. Dr. Steinmetz had regarded these two conceptions as inseparable and said that we should not be disturbed about being unbelievers but for immortality, which needed a God to give it reality.

"On practical religion, especially as it affected Christianity, Steinmetz was equally emphatic. His views are all too easily taken as merely destructive. As a matter of fact, he had not only a philosophy of life, but a support in sterling character which convinces all who knew him that such a life is desirable in all men.

"One afternoon, as we two sat talking about conditions in the world of human relations, the doctor said to me, 'No human being should engage in an unsocial act.' I then asked him what objection there could be in tying up more closely with Christianity as a practical means of redeeming affairs. With pity in his eyes, he said, 'Christianity has failed; anything that has been advocated for two thousand years, and has been no more successful than for men to hate each other so that the biggest war in history has but recently closed, has failed.'

"Dr. Steinmetz then hastened to explain that he had no objection to the fundamental ethical and spiritual teachings of Jesus, although he considered some of them too involved for a world like this. But he was satisfied that too much was left to incompetency under Christianity. The point of this contention was that man trusted too much to good intentions, aspirations, and so forth, whereas adequate knowledge and proper equipment were necessary.

"And yet Dr. Steinmetz took satisfaction in being a church-member. His attendance was largely confined to occasions when the children were prominent in the service, for his affection for them was unbounded. But his actual connection with a church,[1] his preaching the annual sermon for the Laymen's League of that church, as well as other interests which he from time to time manifested, showed that in him was no violent antagonism toward the church as a whole.

"He died as he lived—a simple, whole-hearted and devoted servant of humanity."

Too broad-minded to ignore the influence of religion in the lives of many of those about him, and in the world in general, Dr. Steinmetz recognized religion as a definite characteristic of the human race. What particularly interested him, as was to be expected, was the relation of religion and science in man's practical every-day life.

About a year before he died, Dr. Steinmetz delievered an address in which he summed up his conclusions on this broad subject. It was an address that gave only a hint of his own innermost convictions. But it contained statements in which religionists could and did take comfort, for he frankly pointed

[1] The First Unitarian Society, All Souls' Church, Schenectady, membership in which is entirely without test or requirement of belief.

out that while science returns a purely negative answer to the colossal question as to whether or not there is a God, and whether or not man has an immortal soul, still this negative answer "is not conclusive, and the question is still as open as it ever was." Furthermore, he reiterated a great truism, that religion must be based inevitably upon faith, and in science nothing is accepted on faith; everything must be established by proof.

This opinion of the electrical master was encouraging to the followers of the faith only in a negative sense, to be sure. They were glad to find that Dr. Steinmetz did not ruthlessly rule out of existence both God and immortality; he merely stated that science could not recognize these things, but that this position of science did not settle the question. God and immortality might exist just the same.

It was the conclusion of an agnostic. It was the conclusion which any man who accepted only what science would demonstrate to him and rejected what science could not substantiate must have reached.

The address itself is a clear exposition, touching briefly on the historical setting of the two great central conceptions, defining broadly the goals which both science and religion strive to attain. Uttered by an agnostic of strong scientific leanings, it is unquestionably fair and just in its treatment of the subject which Dr. Steinmetz gave it: "The Place of Religion in Modern Scientific Civilization."

This address was delivered on November 5, 1922,

in All Souls' Church, Dr. Steinmetz occupying the pulpit on that day by special invitation of the minister. The lecture—for such it is much more than a sermon—is given in full, since it reveals the thought of a unique personage on a subject of unique significance.

"Religion," said Dr. Steinmetz, "may be defined as dealing with the relations of man to superior entities, usually conceived as individuals, that is, a personal God or personal Gods, or conceptions of immortality, resurrection, etc. Science, as understood here, deals with the conclusions derived by the laws of logic from our sense perceptions.

"There has grown up through the centuries an increasing antagonism between science and religion, making the two incompatible with each other. Religion met this by the Inquisition, that is, by forcibly suppressing science and its votaries. However, our civilization is an engineering civilization, and the prosperous life of the large populations which our earth now supports has become possible only by the work of the engineer. Engineering is the application of science to the service of man, and so to-day science is the foundation not only of our prosperity, but of our very existence, and thus necessarily has become the dominating power in our human society.

"Therefore, in civilized countries, the attempt of religion to suppress science by the Inquisition of medieval times is laughed out of court, and in semi-civilized countries such an attempt, for instance, as to forbid by legislation the teaching of evolution in state universities, is rarely successful, at least not for a long time.

"Our knowledge of the superior entities, with which historic religion deals, has been derived by experience and by 'revelation.'

"Undoubtedly experience led to the first conception of superior beings, or 'gods': the forces of nature personified; the experiences in dreams; the orderly progress of nature, which seemed to imply a manager of the universe. With our increasing knowledge, this became less and less satisfactory. For instance, the terror of the thunder-storm led primitive man to the conception of a Supreme Being, whose attribute was the thunderbolt. But when Franklin brought the lightning from the clouds and showed it to be a mere electric spark, when we learned to make lightning harmless by the lightning-rod, and when finally we harnessed electricity to do our work, naturally our reverence for the thrower of the thunderbolt decayed.

"So the gods of experience vanished. For a time, the wonderful fitness of nature gave argument for the defense of the conception of a Supreme Being who had made everything in nature so perfectly fitting its purpose. But Darwin gave a ridiculously simple explanation of the fitness of nature, by showing that only the fittest can survive and anything unfit is rapidly exterminated, and so, on merely mechanical principles, nature becomes perfect in fitness.

Therefore, the hatred of all the forces of darkness against the theory of evolution.

"Thus no evidence or proof of the existence of a God has been found in the phenomena of nature, based on experience.

"Logical proof of the existence of God, immortality, etc., has been given repeatedly, but in all these attempts, the proof is based on an assumption, such as the existence of the mind independent of the body, etc., and stands and falls with this assumption, and this assumption cannot be proved. That is, the attempted proof of God is a syllogism. There are a number of conceptions, such as God, immortality, the existence, independent of the body, of the soul and the mind, and in general, the personal individuality or the ego, the existence of life not merely as a biochemical process, but as an entity independent of the living body, etc. All these conceptions are interrelated and dependent on each other in such manner that, if one is proved, all the others logically follow from it.

"For a long time attempts were thus made by empirical science to prove one of these conceptions. The most promising appeared to be the proof of the existence of life independent of and not limited by the laws of inanimate nature. Such proof would be brought if the existence of a 'vital force' could be proved, which can do things that the laws of inanimate nature cannot accomplish. As you know, the vital force conception held sway for a long time in chemistry. When chemistry developed, it was found that the compounds met with in inanimate nature could be produced by the chemist in his laboratory; but the 'organic' compounds, that is, the things produced in living plants and animals, could not be produced by the chemist, and so were attributed to the action of the vital force. But gradually one after the other of these chemical products of living things were produced synthetically by the chemist, and while many of the organic compounds have for some reason or other not yet been reproduced, the evidence has long become conclusive that the same laws of chemistry operate in the construction of the organic compounds, and no field is left for a vital force, but the vital force does not exist in the realm of science.

"In the realm of science, all attempts to find any evidence of supernatural beings, of metaphysical conceptions, as God, immortality, infinity, etc., thus have failed, and if we are honest, we must confess that in science there exists no God, no immortality, no soul or mind as distinct from the body, but scientifically God and immortality are illogical conceptions. That is, science had inevitably to become atheistic.

"There remained only revelation as the foundation of the historical conception of religion. But is there any difference between the 'dream' of prehistoric man, in which he 'sees' wonderful things, and the 'revelation' of Mohammed or Buddha or Moses or other founders of religions, which all contradict each other?

Steinmetz exchanging experiences with Elbert Hubbard at
Association Island, Lake Ontario, in 1913

"Thus there is no evidence outside of science for God, immortality and similar conceptions, and there is evidence against these concertions in science, and science has justified its methods and conclusions by the work it has accomplished.

"But it is very hard for man to get along without a belief in these conceptions. We may get along without a God, but not without immortality. Our self-conceit dislikes to place so little value on ourselves, our knowledge, skill, experience, in short, our ego, to concede that all this is merely a function of the biochemical process of life, which utterly ceases and vanishes with the disintegration of the protoplasm of our body by death.

"The conceptions of physical science are incompatible with the metaphysical conceptions of God, immortality, infinity, etc. But are the conceptions of science really final and all-embracing, or are they limited also, holding within a certain range only, and not beyond this? Science derives its conclusions by the laws of logic from our sense-perceptions. Thus it does not deal with the real world, of which we know nothing, but with the world as it appears to our senses. But are there no limitations to our sense-perceptions, which limit the validity of the conclusions we derive from them?

"All our sense-perceptions are limited by, and attached to, the conceptions of time and space. Kant, the greatest and most critical of all philosophers, denies that time and space are the product of experience, but shows them to be categories, conceptions in which our mind clothes the sense perceptions. Modern physics has come to the same conclusion in the relativity theory, that absolute space and absolute time have no existence, but time and space exist only as far as things or events fill them, that is, are forms of sense-perception.

"Still greater and more pertinent is another limitation of our sense-perceptions: our senses can perceive only finite things, but cannot perceive the infinite. No reasoning from any foundation can put anything into the conclusions which is not contained in the foundations, and thus, with our sense-perceptions finite, all conclusions from them, that is, the entire structure of science, are limited to the finite. Hence any attempt of science to deal with an infinite conception, as the infinite in time and space, immortality, and the conception of God, etc., in short, all those interrelated conceptions already discussed, must fail and lead to contradictions, and show these conceptions as illogical.

"Thus the proof of the non-existence, in science, of the conception of God, immortality, etc., really means nothing except that we cannot get by reasoning a conclusion which is not contained in the premises on which we started our reasoning; finite science cannot deal with the conceptions of the infinite or absolute.

"Furthermore, science derived its conclusions from the sense-perceptions by the laws of logic. But what proof is there of the correctness of the laws of logic, except experience, which, no matter how comprehensive, must remain limited?

"Thus the negative answer of science on the question of whether there are conceptive entities of infinite character, as infinity in time and space, immortality of the ego, God, etc., is not conclusive, and the question is still as open as it ever was.

"How can we approach its solution, and can we ever get an answer on the question of the existence of the infinite? The best we can expect to do is to search into the foundations and limitations of our mental processes, to determine how far conceptions are really illogical and contradictory, and how far they appear so, merely because they involve conceptions beyond the limits of our mind. There can be no scientific foundation of religion. Belief must always remain the foundation of religion, while that of science is logical reasoning from facts, that is, sense-perceptions. All that we can say is that the two, science and religion, are not necessarily incompatible, but are different and unrelated activities of the human mind.

"Thus, inherently, science and religion are not antagonistic, but separate, the one dealing with the finite conclusions from our finite sense-perceptions, that is, the world of facts and reality, and the other with infinite conceptions, which can neither be proved nor disproved empirically, but are outside of the realm of science, in the field of belief. A collision between science and religion can occur only when the one tries to encroach on the field of the other, as, for instance, when religion attempts to teach history in the fable of the creation of the world, or biology in opposing the theory of evolution, etc., These are not proper and essential parts of religion. They are mere survivals of the child's age of man."

CHAPTER XXV

THE GREAT TORCH SUDDENLY EXTINGUISHED

THE year 1923, in the life of Dr. Steinmetz, gave no promise at the outset of the events that were so near. The doctor was quietly busy with all his accustomed activities through the latter part of the winter and into the spring and summer.

His one absorbing line of investigation was his general study of transient phenomena, particularly an intensive series of experiments with his artificial lightning. In these experiments he sought no spectacular effects but perseveringly worked for results which seemed intangible and obscure to all but his technically trained associates.

During the winter he improved his lightning-generator. With the new apparatus he embarked upon research even more interesting to a technical man than those that had startled the public a year or so previously.

Gradually, however, the need for increased facilities in this work was made apparent. The scope of the investigations was steadily broadening, and the electrical capacity of the apparatus was increased in proportion. By the spring of 1923 Dr. Steinmetz and his co-workers had planned sufficient experimental work, as a result of what they had discovered up to that time, to keep them busy for three years to come.

The nature of the laboratory work had made it evident, by this time, that special quarters would be desirable. A third new lightning-generator had just been completed, under the direction of Dr. Steinmetz, having a capacity for producing more powerful artificial lightning than either of its predecessors. It came closer still to imitating nature's lightning, although it was still a puny exhibition in comparison with the grandly awesome product of an actual electrical storm.

The first experiments with this apparatus, and indeed some of the experiments with the smaller machine previously in use, caused a considerable upset in the electrical circuits of the building. Five hundred thousand volts, and close to two million horse-power, was unloosed in a single flash of laboratory lightning, affecting other electrical equipment rather unexpectedly.

As erratic as the terror of the skies itself, the discharge of this miniature lightning fairly disrupted some electrical apparatus. The high voltage backed up in the main circuit of the laboratory building, and on the occasion of one such an outbreak it sent a wave of electrical energy into the wires that supplied an electric glue-pot, used by some carpenters, burning out the glue-pot and compelling the carpenters to quit work.

These difficulties led to the decision to build a special laboratory-room for Dr. Steinmetz. The general engineering laboratory began the construction of such a room in the early summer of 1923, planning to locate it on the roof of one wing of the laboratory building, entirely apart from the rest of the laboratory.

The new lightning-machine was to have been installed in this room. Other electrical equipment was to be added, and the current for operating the lightning-generator was to be supplied by a motor generator set, which would have made the Steinmetz laboratory independent of the main electrical circuit of the building, so far as providing a source of energy for the lightning-generator was concerned.

The actual construction of these new quarters for Dr. Steinmetz was well advanced by early autumn, at the time of his death; and he had anticipated with keen delight the advantages to his work which the new facilities would have provided, but of which he was never to make use.

In the spring of 1923, Dr. Steinmetz remarked to the Haydens that during the summer he wanted to carry out a plan he had been pondering for some time. This proved to be an excursion to the Pacific Coast, a region of the country that he had never seen.

CHARLES PROTEUS STEINMETZ

The idea was talked over for a long while; and then came a development that had much to do with crystallizing the matter and with fixing the approximate date. The Pacific Coast section of the American Institute of Electrical Engineers invited Dr. Steinmetz to be one of its speakers at its annual autumn meeting, to be held October 3 and 4, at Del Monte, California.

This invitation fitted in very well with the doctor's general desire; and it was finally made the nucleus around which was laid out the program of his memorable western trip; memorable because of the remarkable reception accorded him by the public, with the sad sequel that came just two weeks after his return to Schenectady.

The trip was the all-absorbing thought in the doctor's mind through the spring and summer. But it was not allowed to interfere with a family pilgrimage of the Haydens, which occurred every year on Memorial day, and in which Dr. Steinmetz always shared with hearty accord. Mr. and Mrs. Hayden invariably took advantage of this holiday to pay a visit to the original home of the Hayden family, at Haydenville, Massachusetts. There, where the past generations of the line slept in the village cemetery, it was the custom to hold a family reunion. Dr. Steinmetz never more completely exhibited his unity of life with his adopted son and grandchildren than on these occasions. He was one of them in every respect, and took part with the children in their games and revels.

Through the summer, preparations for the western tour of Dr. Steinmetz and his adopted family went forward steadily. The entire family made ready to go, Mr. and Mrs. Hayden and the three children, Joe, Marjorie, and Billy. Dr. Steinmetz would not have enjoyed the journey half as much if any one in the household had been unable to participate.

He expressed a particular desire to visit the Grand Cañon of Colorado; hence, the outward route was arranged to include that region, with Denver as the first stopping-point. Gradually, as the news of his coming got around among western cities, many requests were received at the Wendell Avenue home for luncheon talks or evening addresses by the doctor. Electrical men throughout the West expressed an interest in the tour; and by the time the group left Schenectady, a number of public appearances by Dr. Steinmetz had been definitely agreed upon. These included addresses before many local sections of the American Institute of Electrical Engineers, which hastened to show their esteem for their former national president.

On September 1 Dr. Steinmetz, with the Hayden family, started from Schenectady on the first stage of this trip. They reached Denver on September 3 and remained there until the fifth. In that city Dr. Steinmetz gave his first public address of the tour, an evening address, for which the local committee had engaged a hall of moderate size, seating perhaps nine hundred persons.

Public interest, however, was much greater than even Dr. Steinmetz himself supposed, and it showed itself when the time for the address arrived. Long before the hour set for the doctor's appearance, the hall was overflowing with Steinmetz admirers, and a large throng had gathered outside. Those in charge hastily transferred the assemblage to the largest hall in the city, an imposing auditorium seating nearly five thousand. That, too, was soon filled, so that when Dr. Steinmetz finally walked out upon the platform, he found every seat occupied and many people standing in the aisles and at the rear of the hall. The people's greeting to the electrical master was enthusiastic and mingled with much sincere admiration. His subject on this occasion was "The Electric Power Industry," which was his topic at virtually all of his evening appearances. At noonday luncheons he usually spoke either on "Electricity and Civilization" or "Engineering Problems of the Future."

The scenes at Denver made it apparent that Dr. Steinmetz was widely known by reputation and that people of all walks of life were eager to see and hear him, even though many of them were unfamiliar with his field of work. Nor were the throngs merely curious, for his remarks were always closely followed and not infrequently applauded. And this was but the first of many such experiences throughout the journey.

From Denver the party went on to Colorado Springs, where they visited the Garden of the Gods, the Cave of the Winds, Cheyenne Cañon, and Seven Falls, and rode by automobile over the Hermit Rim drive and the Grand View drive. And then they continued, on September 16, to the Grand Cañon and Los Angeles, where there were more addresses, both at noon and in the evening, including a dinner at which Dr. Steinmetz was the guest of honor.

Part of the itinerary at this stage of the tour included a visit to Hollywood and the principal moving-picture studios. It was here that Dr. Steinmetz was welcomed by Douglas Fairbanks, who showed him through the vast array of stage-sets and explained everything that Dr. Steinmetz wanted to know.

"A grown-up boy, with all the enthusiasm of youth, and all youth's unquenchable thirst for knowledge, that was Charles Proteus Steinmetz as he revealed himself to me on his visit to Hollywood," said Douglas Fairbanks, describing the unique occasion.

"He was tremendously interested in everything pertaining to studio life and expressed surprise at the infinite amount of detail connected with the producing of a large motion-picture. Of course, he went into everything which seemed possessed of any scientific angle whatever.

"He asked a thousand questions. His mind

seemed to assimilate ideas almost before they were expressed. His enthusiasm and his vigor were really contagious. His eyes were mere agents of his brain. The minute he looked, a question formed on his lips."

This is a fleeting picture of Dr. Steinmetz in a period of complete relaxation, out for a holiday, yet a holiday mixed with much that was profitable. He enjoyed this Hollywood experience; still, his greatest pleasure as the trip proceeded was the panorama of matchless Western scenery which was unfolded before him. He would stand for a long time gazing at a great plunging waterfall, or a tall, silver-like thread of a stream plunging down in a narrow cataract. And it was the esthetic appeal that had touched him at these moments. Let it not be supposed that in his scientific mind he was picturing these streams harnessed to a power-house and giving of their energy to serve men electrically, while surrendering their ability to thrill men's love for the beautiful. He thought not of this, or if he did think of it, the thought was quite secondary in character. He was the ardent, deeply stirred lover of nature throughout this visit to scenes of nature's wonderland.

This was his reaction, too, at night, when the luxurious Pullman train was speeding onward through regions of wild appeal. Once or twice Mr. Hayden caught a glimpse of him on the sleeper. He was in his berth but not sleeping. Instead, he was crouched on hands and knees, gazing out of the window at the

country-side, passing in quick panorama under the
sheen of the moon. It seemed as if he could hardly
bear to cease contemplation of the boundless expanse of Western lands and the picturesque vistas
that continually appeared, not even long enough to
sleep.

On September 23 the party reached San Francisco,
where there was another round of luncheon talks, a
technical lecture in the evening before the local section of the American Institute of Electrical Engineers, and much enjoyable sight-seeing. A week later
the travelers proceeded to Del Monte, to attend the
Pacific Coast convention of the institute.

There Dr. Steinmetz delivered a formal paper,
constituting his most technical address of the trip.
Its subject was "High Voltage Insulation." Mr.
Hayden was co-author with Dr. Steinmetz in this
discussion, which was received by the convention with
keen interest. Its importance lay in the way it dealt
with the most recent investigations of Dr. Steinmetz
into one phase of the general question of improving
electrical transmission so that even greater progress
can be made in the future than has been accomplished
in the past.

The return-trip to Schenectady began on October 6
and lay through Salt Lake City, where there was another series of talks. Again the evidence was present of great popular interest in the "little giant" of
the electrical world. The entire party was present

on one of these days, at an organ-recital in the Mormon Tabernacle.

On the way from Salt Lake City to Chicago, Dr. Steinmetz met with one of the picturesque experiences of his later life and was one of the principals in a peculiarly interesting episode. This was his friendly meeting with William Jennings Bryan, who chanced to be traveling on the same train.

Upon learning that Dr. Steinmetz was on the train, Mr. Bryan inquired the way to his state-room, where he was very cordially greeted. The two men, each of public prominence in widely different spheres, engaged in conversation for almost two hours. And, as might have been expected, the dominant topic was that of the relation between science and religion.

Two men with more sharply different views on this question could hardly have come together. It was the meeting of a Christian believer of the orthodox school with an avowed agnostic, whom some thought to be only a few degrees removed from a state of learned atheism.

They proceeded at once to sound each other out. In reality, Mr. Bryan was perhaps the most inquisitive of the two. He was eager to know just what Dr. Steinmetz's views were and asked several direct questions, which Dr. Steinmetz answered with cordial courtesy and evident sincerity.

He told Mr. Bryan, in substance, that he could not believe anything he could not prove. His scientific training and scientific mode of thinking had made such an attitude essential, in his opinion. He could and did believe all that science proved; but science could not prove the principles of religion, the reality of God, of the soul, of immortality; things which, he pointed out, require faith in order to accept. He could not personally accept anything on faith and hence could not accept religion.

Yet he made it clear to Mr. Bryan, as he had made it clear many months before to a church congregation in Schenectady, that science and religion need not conflict. Each had its own spheres of activity and influence.

Dr. Steinmetz, in brief, frankly declared himself an unbeliever. But he indicated his broad point of view when he remarked that he believed, without the least reservation, in having the Bible taught to children. He thought it would be beneficial to them, and that only good reactions could result from inducing them to know it as it stands.

This interesting visit of the famous politician, lecturer, publicist, and Christian leader with the master mathematician, electrical engineer, physicist, and eminent scientist concluded some time before Chicago was reached. The two parted with as much cordial feeling as they had shown upon meeting, neither seemingly any the less friendly because they held such divergent opinions on matters commonly admitted to be at the base of human welfare.

Dr. Steinmetz, whatever his religious views, was

warm-hearted, sympathetic, and considerate of other people's feelings every moment of his life. He indicated this in a rather striking way shortly before their journey's end. The Hayden children, to while away a few moments, had begun reciting a childish jingle and invited the doctor to join with them. But the jingle contained reference to an individual of the negro race; and near them in the Pullman car the negro porter was at work, well within ear-shot. Dr. Steinmetz, perceiving this, could not be induced to join in reciting the jingle. He would not run the risk of having the porter overhear them and feel humiliated.

At Chicago, as elsewhere on the trip, the doctor showed some dislike for walking. Frequently he had been obliged to walk something of a distance from train to train, or from train to automobile. These moments, even though brief, seemed to him to be the most irksome of the entire journey.

At length, on October 12, their train rolled out of the West and stopped at the familiar Schenectady station. Half an hour later they entered the home on Wendell Avenue, and the six weeks trip was at an end.

As he walked into the reception-hall of the house that had come to be a very happy spot to him, Dr. Steinmetz remarked to Mr. and Mrs. Hayden: "I am glad to get home again; and I am glad we took the trip. I think it was worth while." He had previously told them, while on the train to Chicago, that

One of the last photographs of Dr. Steinmetz. He is in the moving-picture colony at Hollywood, California, and with him is Douglas Fairbanks

he had had a fine time and had enjoyed the trip to the utmost extent.

"Our next trip," he said, half jokingly, yet seriously, too, "will be to the Mediterranean. That will not be so tiring, perhaps. It will be an ocean trip, with only two weeks on land, and then the return."

At no time on the trip had he said anything about feeling unwell. He got tired at times; and in one city he underwent a medical examination, chiefly at Mr. Hayden's behest. The physician who made the examination told Mr. Hayden that Dr. Steinmetz's heart was functioning poorly, but this was not disclosed to Dr. Steinmetz.

When he got home he had his personal physician look him over, not because he was in any sense fearful, but as a methodical procedure. His physician thought he needed a good rest and told him to stay in bed for a while. But Dr. Steinmetz did not enjoy that. He was unwilling to be in bed in the daytime. He gave his family some difficulty keeping him quiet and warm, insisting as he did on getting up at different times through the day. However, he did consent, after a while, to stay on the second floor, in order to avoid the tiring exertion of going up and down stairs. He was only down-stairs once after he got home; every day, with that one exception, he had his meals in his room, and nearly always had his breakfast before arising.

Time passed tediously with him under the rest treatment. Although always fond of reading, particularly relishing the mystery type of story in moments of relaxation, he found it hard to keep his mind occupied during these last days. He had read, with much delight, the entire series of Tarzan stories, and had reveled in all the most thrilling detective stories—always avoiding, in all his fiction reading, those stories which he heard had unhappy endings, and invariably expressing regret if he stumbled upon such a narrative.

Apparently he had no premonition that the end was near. He displayed no trace of apprehension, much less of panic, although there is reason to believe that he had for some time understood the condition of his heart. To the very last, however, he was mentally tranquil, calm, and well poised, with his thoughts busily occupied.

The night before his death he was up and dressed for a while, although he did not venture to come down-stairs. He spent part of the evening reading, finding some passages of interest in Humphrey's "Physics of the Air." Some of the phenomena which this book mentioned, and one or two of the illustrations of devices for recording changes in the atmosphere, recalled what he and Hayden had witnessed, during their trip, at the Lick Observatory. He marked some of these passages with a pencil, to call Hayden's attention to them. He also read a magazine fiction story.

Shortly after 9:30 o'clock he retired, with his usual cheery good night to the family.

THE GREAT TORCH GOES OUT

About eight o'clock on the morning of October 26, 1923, Mr. Hayden went up to Dr. Steinmetz's room to see how he had passed the night. He found the doctor awake and comfortable but somewhat restless. They conversed for a few moments, and Mr. Hayden insisted that the doctor take a dose of medicine. Then he told the doctor he would have breakfast sent up to him, and that it would be better not to try to get up until he had eaten something.

"All right," agreed Dr. Steinmetz. "I will lie down."

Those were his last words. Mr. Hayden left the room and on the stairs passed his son, Billy Hayden, coming up with a breakfast-tray prepared for Dr. Steinmetz. The boy entered the doctor's bedroom and announced breakfast. There was no reply from the bed. Drawing closer, Billy noticed that Dr. Steinmetz seemed very still, and his anxiety was aroused. He hurried down-stairs and called his father. Mr. Hayden at once telephoned for the physician, but there was nothing whatever that human aid could do. Dr. Steinmetz had died quietly, painlessly, all in a minute, between the time that Mr. Hayden left his bedside and the time when young Billy Hayden entered the room.

The sad news spread through the city like an electric flash. Within a few moments it was known at the General Electric plant, and in an unbelievably short interval it had spread from end to end of the big establishment. Officials of the company were soon at the house on Wendell Avenue, to offer condolences and express the tribute of the organization through which Dr. Steinmetz had served mankind.

All that day and the next day, telegrams came to the General Electric Company from individuals, scientific bodies, electrical concerns, throughout the United States and from abroad, communicating the universal regret which the loss of the electrical master aroused, and expressing the admiration which all men felt for his undisputed genius.

Two days later the public of Schenectady, people of high and low degree, representative of the rank and file of folk everywhere who have benefited indirectly, yet definitely, from the work of Dr. Steinmetz, paid their own homage at his bier. The great master of electricity lay in state in the home which he had built and loved, while for hours on that dull, drizzly October Sunday an endless line of mourners filed past. Crysanthemums, roses, and those orchids which he particularly loved banked the casket on every side; and the sorrowing continued on the following day when the funeral was held, thousands watching at the grave as the last rites were performed and those luxuriant flowers were laid in a fragrant pile upon the little mound of cold brown earth.

Eminent men served as the honorary pall-bearers: Owen D. Young, chairman of the board of directors of the General Electric Company; Gerard Swope, president of the General Electric Company; Francis C. Pratt, one of the vice-presidents of the company;

THE GREAT TORCH GOES OUT

Dr. Charles A. Richmond, president of Union College; Dr. Ernest J. Berg, consulting engineer of the General Electric Company, and Dr. Steinmetz's house comrade of many years before; and Lieutenant-Governor George R. Lunn, in whose city administration Dr. Steinmetz served in his only period of public life.

The wide-spread regret at the passing of Charles Proteus Steinmetz was accompanied by a momentary interval of surprise when it became known that he left no great estate. His Wendell Avenue house, which in his will he expressly left to Mr. Hayden, had been deeded over to the latter by Dr. Steinmetz six months previously, so that this provision of his will was already fulfilled. Beyond this, he established a trust fund for one of his sisters, Miss Clara Steinmetz, who was living in New York at the time of his death, and remembered with small sums others among his relatives and intimate friends.

But there was no accumulation of riches, such as many persons supposed would prove to be the case, taking into account the eminence of Dr. Steinmetz in his profession and his years of distinguished service with the General Electric Company. Dr. Steinmetz did not seek wealth; nor did he believe in hoarding it. His riches were of a different sort. They were represented—yet only in part—by such honors as the award to him in 1913 of the Elliott Cresson gold medal "in recognition of the successful application of analytical methods to the solution of numerous problems of first practical importance in the field of electrical engineering." They were embodied, these riches of Steinmetz, even more enduringly in the hearts of his fellow-mortals, and in the lives of all who live in these modern days with their widening usage of electrical methods, so many of them depending upon his achievements.

What the estate of Dr. Steinmetz really signified was impressively pointed out in an article in the "Christian Century," quoted by the "Literary Digest" a few months later. The article commented on the persevering efforts of Andrew Carnegie and John D. Rockefeller to give away their millions before they had the opportunity taken from their hands.

"But Charles Proteus Steinmetz, who was one of the world's greatest mathematicians and electrical engineers when he died," said the article, "brought us a new suggestion. He had ample opportunity to accumulate wealth, but he proposed that no man take from society more than he requires for his immediate needs. Let a man house and clothe and feed himself and his family; let him take enough to afford him opportunity for unhampered service to the community. For the rest, let the man take no thought; certainly never let him waste himself in piling up riches. What flows out above his actual needs, let it go back immediately and directly to where it will produce for the public good. This, says 'The Christian Century,' sounds almost like the revolutionary proposals that a crowd of peasants, sitting on a Galilean hillside, heard one day."

In similar vein, a newspaper editor in the neighboring city of Albany, said: "The man who was universally known as the Wizard of Schenectady is dead, and therefore the crowding achievements of a life that was short in proportion to its accomplishments are marshaled in review. They are his estate more truly than the money which he had made, and for which he never greatly cared. They are the heritage he leaves, and not the children of his adoption only, but all the generations born and yet unborn are the heirs of his genius.

"Yet it was not with the things he had done that Dr. Charles Proteus Steinmetz was concerned. His life was always too busy to spend even thought upon the past, unless he saw in the vision that grew larger every year he lived new means of turning old ideas to account. He lived and worked in the future, and the secrets which the future still holds will be his monument even beyond the marvels which he wrought for the present enjoyment and profit of his fellows.

". . . Steinmetz was one of a splendid company recruited by the centuries. Men in many walks of life have taken their places in its ranks, which are never very full. It is the company of the great-hearted and high-souled who spend themselves upon their race. . . . He was beyond all things else, at the end as well as at the beginning, a lover of his kind, who turned his love to practical things."

CHARLES PROTEUS STEINMETZ

The great tide of honorable acclaim and spontaneous appraisal of his life and work was swelled by countless individual tributes. Among the most impressive were those of Herbert Hoover, Secretary of commerce; E. W. Rice, Jr., honorary chairman of the board of directors of the General Electric Company; Professor Harris J. Ryan, president of the American Institute of Electrical Engineers; A. E. Kennelly, professor of electrical engineering at Harvard; Charles F. Scott, professor of electrical engineering at Yale; S. W. Stratton, president of the Massachusetts Institute of Technology; and R. A. Milliken, director of physics at the California Institute of Technology.

This particularly unique paragraph appeared in the "New York Times":

"To describe the freedom given to Steinmetz by the General Electric Company, his friend, Prof. Vladimir Karapetoff, of the Cornell School of Electrical Engineering, has evolved a happy phrase— 'he was allowed to try to generate electricity out of the square root of minus one.' That, doubtless, was what the man often seemed to be doing to those to whom mathematics as he knew it was equally incomprehensible and useless. Fortunately his employers —no genius ever had better and few as good—took a different view."

Striking among the many estimates of his character which his death called forth is the tribute to "Daddy"

Steinmetz which came from Mrs. J. LeRoy Hayden, wife of his adopted son.

"When 'Daddy' Steinmetz passed away," said Mrs. Hayden, "one message we received was 'a tribute of respect to the memory of the brilliant intellect, the big and loving heart and the white soul of Dr. Steinmetz.'

"To us who had called him 'Daddy' for twenty-one years, it was a perfect picture of the man we had known and loved so well.

"He was essentially a home and family man. To be called 'Grand-daddy' meant more to him than to be known as the electrical wizard. To him, children came first. He believed that every child should have everything possible that he or she could want. In return he expected respect and obedience. He, himself, was never too tired nor too busy to answer their questions and seemed happiest when included in their plans. He liked to be one of them.

"He was most hospitable and always enjoyed guests. While he did not play cards, he loved a game of chess. The one friend with whom he played was more often victor than loser. Daddy took his defeat good-naturedly and said he'd do better next time.

"He was never critical of others, never thought ill of any one, but was always willing to give the benefit of the doubt.

"Chapters have been written of his greatness intellectually; as many more could be filled with his generosities and kindnesses to others. Dwarfed, perhaps, in body but with a heart as big as the universe and a soul as white as a child's."

There were many memorial gatherings in the few days immediately following Dr. Steinmetz's death. One of the most impressive was held at All Souls' Unitarian Church, in Schenectady, from the pulpit of which, a year previously, Dr. Steinmetz had enunciated his views on the relation between science and religion. The principal speaker was E. W. Rice, Jr., the man who, thirty years before, had arranged with Steinmetz to become an employee of the newly organized General Electric Company. Mr. Rice, during his remarks, related a conversation among some neighbors, who were mourning the death of the master of electricity.

One of them exclaimed, with deep emotion, "And he was my friend."

His small son of seven years looked up from his play and added, "He was my friend, too, daddy."

And that is how he will be remembered by many, next after the memory of his engineering brilliance— as a "friend of man."

INDEX

INDEX

INDEX

INDEX

INDEX

INDEX

INDEX

www.ingramcontent.com/pod-product-compliance
Lightning Source LLC
Chambersburg PA
CBHW081049280326
41928CB00053B/2895